A STRUCTURES PRIMER

Harry F. Kaufman, PE, NCARB

Freehand Graphics by

Howard F. Itzkowitz, RA

Prentice Hall

Upper Saddle River, New Jersey
Columbus, Ohio

Library of Congress Cataloging-in-Publication Data

Kaufman, Harry F.
 A structures primer / Harry F. Kaufman; freehand graphics by Howard F. Itzkowitz. — 1st ed.
   p. cm.
Includes bibliographical references.
ISBN 978-0-13-230256-2 (pbk.)
1. Structural analysis (Engineering) I. Title.
TA645.K377 2010
624.1—dc22

                          2008030343

Vice President and Executive Publisher: Vernon R. Anthony
Acquisitions Editor: Eric Krassow
Development Editor: Dan Trudden
Editorial Assistant: Sonya Kottcamp
Production Coordination: Kelly Keeler, GGS Higher Education Resources, a division of PreMediaGlobal, Inc.
Project Manager: Maren L. Miller
AV Project Manager: Janet Portisch
Operations Specialist: Laura Weaver
Art Director: Jayne Conte
Cover Designer: Bruce Kenselaar
Cover Image: Harry F. Kaufman and Howard F. Itzkowitz
Director of Marketing: David Gesell
Executive Marketing Manager: Derril Trakalo
Senior Marketing Coordinator: Alicia Wozniak

This book was set in Palatino by GGS/PMG and was printed and bound by Hamilton Printing Co. The cover was printed by Phoenix Color Corp.

Pearson Prentice Hall™ is a trademark of Pearson Education, Inc.
Pearson® is a registered trademark of Pearson plc
Prentice Hall® is a registered trademark of Pearson Education, Inc.

Pearson Education Ltd., London
Pearson Education Singapore Pte. Ltd.
Pearson Education Canada, Inc.
Pearson Education—Japan

Pearson Education Australia Pty. Limited
Pearson Education North Asia Ltd. , Hong Kong
Pearson Educación de Mexico, S.A. de C.V.
Pearson Education Malaysia Pte. Ltd.

Prentice Hall
is an imprint of

www.pearsonhighered.com

10 9 8 7 6 5 4 3 2 1
ISBN-13: 978-0-13-230256-2
ISBN-10:   0-13-230256-X

Denise,
For all your help and encouragement

PREFACE

Only after practicing as a registered architect and a professional engineer for over a decade, did I discover my true vocation—that of a college professor. Teaching the subject of "structures" to students of architecture, construction, and technology was a prefect fit for me. I have to admit, though, that my first group of students were not necessarily thrilled with my teaching techniques or the subject matter. I relied on available texts, which were generally written by and for engineers, and quickly overwhelmed and frustrated my students. As I gained experience, I relied less on the engineering texts, and more on my own notes and handouts to teach building structures. I found that this approach was much better received by the students who began to build a solid foundation in their knowledge of structures, and as a result, more quickly learned the advanced material-specific subjects. I have revised and incorporated into this book many of the lecture notes and problems I used to convince non-engineering students that the subject of "structures" is at least understandable, and ideally, interesting.

From my experience I have found that many non-engineering students dread taking structures courses. I believe that the reason the subject of structures holds such dread for some students is that it is frequently taught by engineers (like me), using engineering texts, at the level of detail needed only by budding engineers. While architects and other professionals certainly need to understand structures, they do not need to be as conversant in the subject as to supplant engineers. Instead, these non-engineering students need to know enough about the subject to understand the important role it plays in the design and construction of buildings and to realize when to seek the assistance of and speak intelligently to the necessary consulting professional engineer. Of course, students do need to know enough about structures to pass the professional licensing exam. This book alone will not help meet those objectives, but it will help students understand the basics and lay the foundation on which students will build a greater understanding of building structures.

One other aspect of this book that may be a bit different from other texts is the type and number of graphics. I have found that my students learn better through graphic representations than through words. While I could not get rid of all of the words, I did graphically illustrate as many of the concepts as possible. My dear friend and colleague, Howard Itzkowitz, is responsible for creating the gorillas and for rendering the freehand-graphics found throughout the text. They are meant to be entertaining and good facilitators in the learning experience.

I hope this text will help students understand elementary structural concepts and make their learning enjoyable. Best to all who use this text.

Instructor Resources

To access supplementary materials online, instructors need to request an instructor access code. Go to www.pearsonhighered.com/irc, where you can register for an instructor access code. Within 48 hours after registering, you will receive a confirming e-mail, including an instructor access code. Once you have received your code, go to the site and log on for full instructions on downloading the materials you wish to use.

ACKNOWLEDGMENTS

I would like to thank the following reviewers for their helpful insight: Alexander Schreyer, University of Massachusetts; Bashar Haddad, Indiana State University; Robert Miller, Clemson University; and Dana K. Gulling, Savannah College of Art & Design. I would also like to acknowledge all of those wonderful engineering and architecture professors who taught me, the reluctant student, the subject of structures. It is through their efforts, and with the assistance of the texts I used throughout my schooling and teaching, that I came to understand the fascinating subject of "structures." It is only through the dedication of those before me that I am now able to write this text.

Harry F. Kaufman

CONTENTS

CHAPTER 1

Forces, Moments, and Loads

INTRODUCTION

The primary purpose of a structure is to transfer all of the building forces, moments, and loads back to the earth. To accomplish this, each structural element (joist, beam, girder, column, etc.) must be resistant and transfer these forces, moments, and loads in a logical and systematic fashion. Before we can design any structural elements, we must have a clear definition of the forces, moments, and loads.

1.1 FORCES

What is a force? In the simplest terms, a *force* is a push or a pull. In Figure 1-1(a), a gorilla is standing on a post. In this case, the weight of the gorilla is the force

(a)   (b)

FIGURE 1-1

pushing down on the post. In Figure 1-1(b), the gorilla is hanging from the vine. In this case, its weight is the force *pulling* on the vine.

In Newtonian terms, a force is a vector that is defined by its magnitude and direction (Figure 1-2). To use the term *vector*, we must first clarify our understanding of magnitude and direction.

FIGURE 1-2

1.1.1 Magnitude

In Figure 1-3, the *magnitude* of the force created by the gorilla is its weight expressed in terms of pounds (lbs), kilopounds (kips), newtons (N), or kilonewtons (kN). The units we select in structural problems will be determined by the given data. If a problem is given in SI (International System) units, we would use newtons or kilonewtons. If the problem is given in English units, we would use pounds (lbs) or kilo (1,000) pounds (kips).

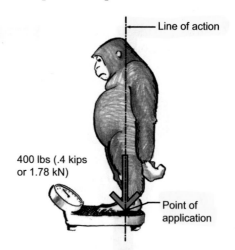

FIGURE 1-3

1.1.2 Direction

In the above example, the direction of the force is vertical, along its *line of action* (Figure 1-1a). Going back to Figure 1-1b, would the force on the vine be any different if the gorilla hangs 2 feet, 8 feet, or 16 feet from the top (Figure 1-4)? Of

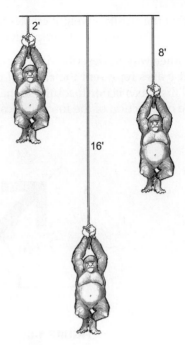

FIGURE 1-4

course, the answer is "No." The line of action of the force created by the gorilla's weight is vertical. We can move that force (the gorilla's weight) anywhere along the line of action and not change the effect that the force has on the object (the vine) on which it acts.

Let us now examine another characteristic, the *sense of direction*. It is insufficient to merely state that the direction of the force is vertical. In Figure 1-5, it is

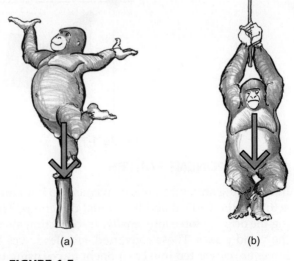

(a)   (b)

FIGURE 1-5

evident that the gorilla's weight creates a force that is downward, a term that provides us with the complete description of direction, vertical and down.

Another way to describe direction is illustrated in Figure 1-6. Assume that the x and y axes represent the cardinal points of the compass (the x axis is east-west and the y axis is north-south) so that the direction is indicated by the *angle* θ. The sense of direction of the force, indicated by the arrowhead, is northwest.

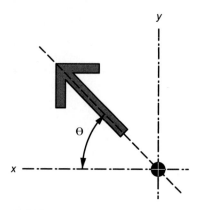

FIGURE 1-6

The direction of a force can also be indicated by a convention commonly used in construction—*slope*. Again, we assume the x and y axes represent the cardinal points of the compass. If the direction of a given force forms the hypotenuse (side opposite the right angle) of a right triangle where the sides parallel to the axes are 9 and 12 as shown in Figure 1-7, the slope of the force is referred to as *9 in 12*. The sense of direction in this case is southeast.

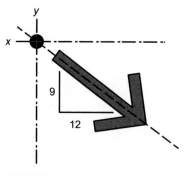

FIGURE 1-7

1.2 COMPONENT FORCES

In designing structures, it is convenient to have all the lines of action of the forces acting in the vertical and horizontal directions. This is accomplished by converting the original force into smaller forces acting along our chosen reference system, the x and y axes. These converted, smaller forces are called *component forces*. The forces are converted into component forces using trigonometry or proportions.

Suppose we have a large gorilla holding a boulder against a vertical surface with a force of 500 pounds at an angle of 60 degrees, as shown in Figure 1-8. The direction of the force is northeast, not directly north-south or east-west or along the x and y axes.

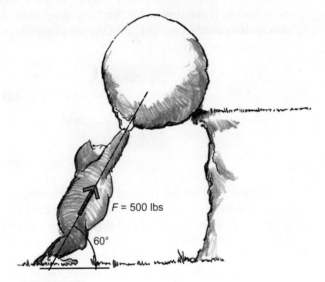

$F = 500$ lbs

$60°$

FIGURE 1-8

We now relieve the large gorilla in Figure 1-8 of its responsibility and replace it with two smaller gorillas that are not quite as strong (Figure 1-9). One gorilla is pulling the boulder with a horizontal force of 250 pounds and the other gorilla is

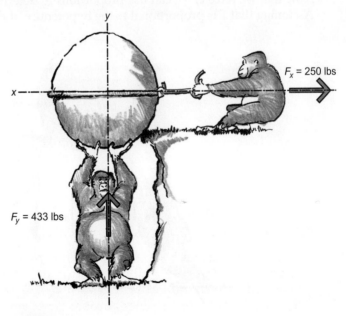

y

$F_x = 250$ lbs

x

$F_y = 433$ lbs

FIGURE 1-9

pushing it vertically with a force of 433 pounds. If the boulder was stable with the force exerted by the large gorilla in Figure 1-8, it will also be stable when acted upon by the two smaller gorillas in Figure 1-9.

Using simple trigonometry we can acquire a better understanding of component forces. If we designate the large gorilla in Figure 1-8 as F, and the small gorilla above pushing north F_y and the small gorilla pulling east F_x, we can develop some simple equations.

$$F_y = F \sin \theta$$
$$F_y = 500 \text{ lbs } (\sin 60°) = 433 \text{ lbs}$$

and

$$F_x = F \cos \theta$$
$$F_x = 500 \text{ lbs } (\cos 60°) = 250 \text{ lbs}$$

Another way we can convert a force into components is by using the slope of the force exerted by the large gorilla. If we assume the large gorilla is pushing the boulder at a slightly different angle, a 4 in 3 slope as shown in Figure 1-10 below, we can utilize this method. When we examine Figure 1-10, the first thing that we will need to calculate the component forces is the value of the c side, or the hypotenuse, of the slope of the triangle. Using the Pythagorean theorem, the hypotenuse c is calculated as follows:

$$c = \sqrt{a^2 + b^2}$$
$$c = \sqrt{4^2 + 3^2} = 5$$

Now that we have c, we can use proportions to determine the component forces. Assuming that F is proportional to the hypotenuse of the slope, or the c side, F_y is

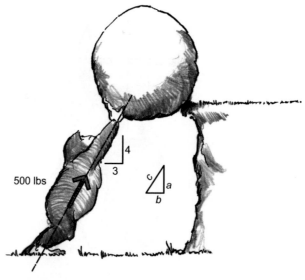

500 lbs

FIGURE 1-10

proportional to the vertical, or a side of the slope and F_x is proportional to the horizontal side or b side of the slope.

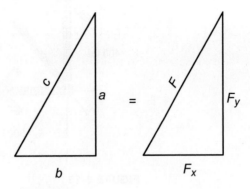

FIGURE 1-11

$$\frac{F}{c} = \frac{F_y}{a} = \frac{F_x}{b}$$

$$F_y = \frac{a}{c}(F) = \frac{4}{5}(500 \text{ lbs}) = 400 \text{ lbs}$$

and

$$F_x = \frac{a}{c}(F) = \frac{3}{5}(500 \text{ lbs}) = 300 \text{ lbs}$$

The two smaller gorillas demonstrate these forces in Figure 1-12.

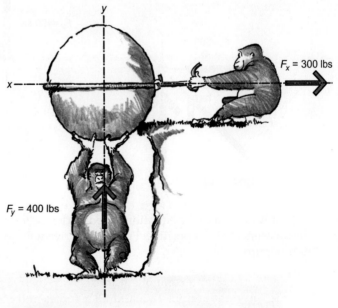

F_x = 300 lbs

F_y = 400 lbs

FIGURE 1-12

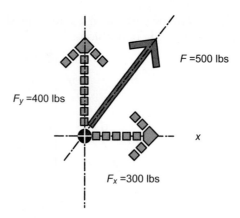

FIGURE 1-13

A more conventional method of diagramming these forces is shown in Figure 1-13. Since F_x and F_y are perpendicular, we refer to these components as *rectangular components*. Rectangular components are the most commonly used components in the study of structures. As mentioned previously, we can move forces (as well as components of forces) anywhere along their line of action. This characteristic is called *transmissibility*. If we move the forces shown in Figure 1-13 along their respective lines of action, the diagram will look like Figure 1-14.

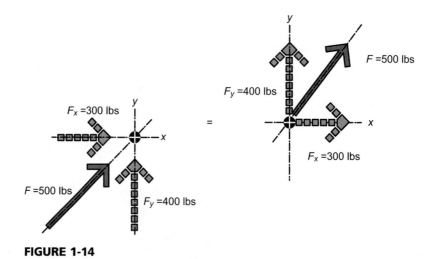

FIGURE 1-14

If we examine the equations used to derive the components, we can see that the answers are both positive. For this text, we will use the following sign convention (Figure 1-15):

- All x forces and component forces with their sense of direction going from left to right are positive and all x forces and component forces with their sense of direction going from right to left are negative.

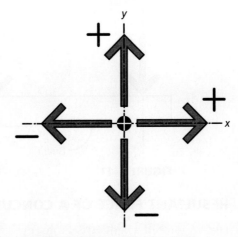

FIGURE 1-15

- All y forces and component forces with their sense of direction going up are positive and all y forces and component forces with their sense of direction going down are negative.

It is important that we stay with this convention throughout the text and exercises.

1.3 FORCE SYSTEMS

Two or more forces make a force system. In a two-dimensional plane there are two force systems. When the lines of action of two or more forces intersect at a common point, as illustrated in Figure 1-16, the system is known as a *concurrent force system*.

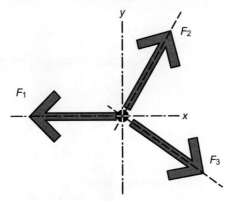

FIGURE 1-16

In a *parallel force system*, demonstrated in Figure 1-17, the lines of action of the forces never meet. Within a two-dimensional plane all force systems are either concurrent or parallel force systems since all nonparallel lines of action will eventually meet.

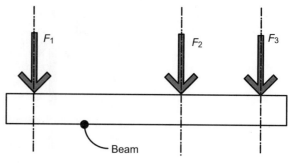

FIGURE 1-17

1.4   RESULTANT FORCE OF A CONCURRENT FORCE SYSTEM

The resultant force of a concurrent force system is simply the opposite of the component forces. The resultant can be substituted for the concurrent force system and has the same effect as the original force system on the object upon which it acts. An example of this is shown in Figure 1-18 below and Figure 1-19 on page 11. In Figure 1-18, one of the two smaller gorillas is pulling the boulder with force of 300 pounds and the other is pushing with a force of 400 pounds to keep the boulder stable. In this case, the large gorilla assumes the role of the resultant and can be substituted for the two smaller gorillas, as shown in Figure 1-19. Under the influence of the resultant force the boulder will remain stable.

It is important to note that the ratio of F_x, the 300-pound horizontal force, to F_y, the 400-pound vertical force, becomes the slope of the resultant. Thus, the resultant is a 500-pound force acting on a slope of 4 in 3 to the northeast.

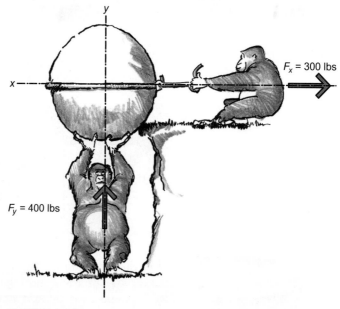

FIGURE 1-18

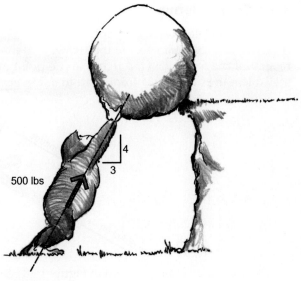

FIGURE 1-19

1.5   DETERMINING THE RESULTANT OF CONCURRENT FORCE SYSTEMS

Three methods are used to determine the resultant of a concurrent force system: the *parallelogram*, the *graphic addition method*, and the *component method*. The parallelogram and the graphic addition method are based on graphics and depend on the accuracy of a drawing. The component method is the most commonly used method by engineers since its precision depends on the use of mathematics.

FIGURE 1-20

1.5.1 Graphic Terms

Both the parallelogram and graphic methods depend on the terms *head* and *tail* (Figure 1-21). The *head* refers to the arrowhead. It provides us with the sense of direction of the force. The *tail* refers to the point of application of the force. The scaled length of the arrow from head to tail indicates the magnitude of the force.

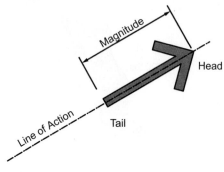

FIGURE 1-21

1.5.2 The Parallelogram Method

To use the parallelogram method we begin by drawing the forces in the system at an appropriate scale to indicate their magnitude and the given direction and sense of direction (Figure 1-22).

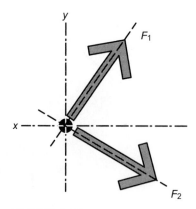

FIGURE 1-22

Once the forces are drawn at their appropriate scale and direction, we draw a line parallel to F_2 at the head of F_1 (Figure 1-23).

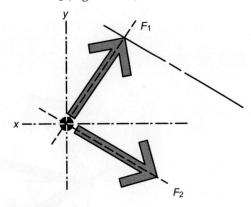

FIGURE 1-23

A line parallel to F_1 is then drawn at the head of F_2, thus forming a parallelogram (Figure 1-24).

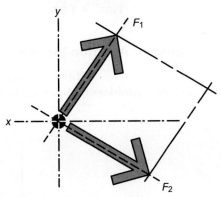

FIGURE 1-24

A third line representing the line of action of the resultant is then drawn from the origin to the intersection of the two parallel lines (Figure 1-25).

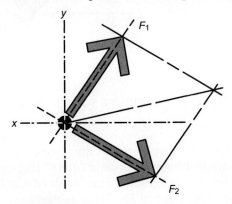

FIGURE 1-25

The head of the resultant will be at the intersection of the parallel lines and the tail at its origin. We can then use a scale to determine the magnitude of the resultant, a protractor to determine the angle of the resultant, ϕ, or drawing instruments to determine the slope of the resultant (Figure 1-26).

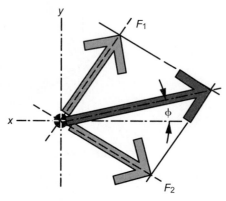

FIGURE 1-26

1.5.3 The Graphic Addition Method

In many ways the graphic addition method resembles the parallelogram method. Let us begin a new problem represented in Figure 1-27.

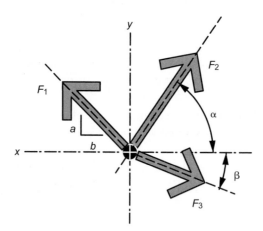

FIGURE 1-27

We begin by drawing one of the forces, F_1, at an appropriate length to indicate its magnitude. Using the slope of "*a* in *b*" (Figure 1-28), force F_2 is then drawn to scale to indicate its magnitude at an angle of α with the horizon. The tail of F_2 must be drawn at the head of F_1 (Figure 1-29).

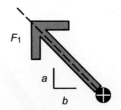

FIGURE 1-28

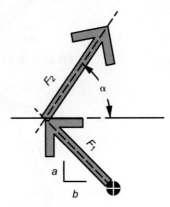

FIGURE 1-29

Using the same process, we continue by plotting F_3 at the established scale at an angle β with the horizon (Figure 1-30). The resultant force is then drawn to close the figure, head to head and tail to tail. The length of the resultant is then scaled to determine its magnitude and the angle ϕ its direction, measured with a protractor (Figure 1-31, page 16). Suppose we take the same forces depicted in Figure 1-27, but draw them in a different order. We see that the magnitude and direction of the resultant force remain constant (Figure 1-32, page 16).

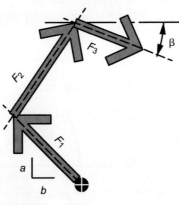

FIGURE 1-30

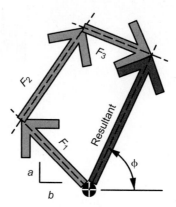

FIGURE 1-31

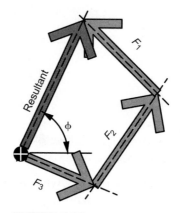

FIGURE 1-32

1.5.4 The Component Method

The most utilized method of determining the resultant force of a concurrent force system involves the use of components. We begin by determining the components of each force in the system. In Figure 1-33, the components can be determined by trigonometry (given the angle) or by proportion (if we know the slope).

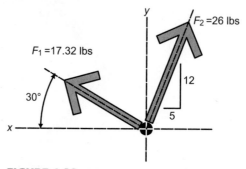

FIGURE 1-33

Given Figure 1-34, we see it is necessary to use trigonometry to determine the component forces of F_1. The x component of F_1, F_{1x}, is calculated by the equation:

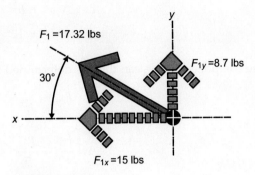

FIGURE 1-34

$$F_{1x} = F_1 \cos 30°$$
$$F_{1x} = (17.32 \text{ lbs}) \cos 30° = -15 \text{ lbs}$$

Note the component F_{1x} is negative because, per our previously established convention, it acts from right to left, so its sense of direction is east to west. The y component, F_{1y} is calculated by the equation:

$$F_{1y} = F_1 \sin 30°$$
$$F_{1y} = (17.32 \text{ lbs}) \sin 30° = 8.7 \text{ lbs}$$

The component F_{1y} is positive since its sense of direction is upward.

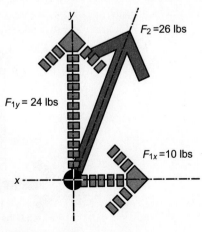

FIGURE 1-35

Since the direction of F_2 (Figure 1-33) is given in slope, it is necessary to first determine the c side, or hypotenuse of the slope triangle.

$$c = \sqrt{a^2 + b^2}$$
$$c = \sqrt{5^2 + 12^2} = 13$$

The components of F_2 can then be determined by using proportions:

$$\frac{F}{c} = \frac{F_{2y}}{a} = \frac{F_{2x}}{b}$$

$$\frac{F}{13} = \frac{F_{2y}}{12} = \frac{F_{2x}}{5}$$

$$F_{2y} = \frac{12}{13}(26\text{ lbs}) = 24\text{ lbs}$$

$$F_{2x} = \frac{5}{13}(26\text{ lbs}) = 10\text{ lbs}$$

Now that we have the components of all the forces in the system, we sum the forces in each axis (indicated by Σ):

$$\Sigma F_x = F_{x1} + F_{x2} = -15\text{ lbs} + 10\text{ lbs} = -5\text{ lbs}$$

$$\text{and}$$

$$\Sigma F_y = F_{y1} + F_{y2} = 23\text{ lbs} + 8.7\text{ lbs} = 32.7\text{ lbs}$$

As shown in Figure 1-36, the sum of forces in the x direction is negative and the sum of forces in the y direction is positive; therefore, the sense of direction for the resultant is northwest from the origin.

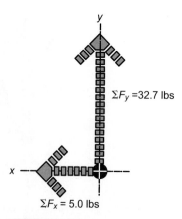

FIGURE 1-36

To determine the direction of the resultant force, we use trigonometry and the sum of forces we calculated above. Since the tangent is a ratio of the sides of a right triangle, we can use the inverse tangent, or arctangent ($\tan^{-1}$), to determine the angle the resultant makes with the x axis.

$$\theta = \tan^{-1}\left(\frac{\Sigma F_y}{\Sigma F_x}\right)$$

$$= \tan^{-1}\left(\frac{32.7 \text{ lbs}}{5 \text{ lbs}}\right) = 81.3°$$

By determining the sense of direction using the signs of the sum of forces, as we did in this problem, we can use the absolute values (ignoring the negative signs) in the θ equation and the calculated angle will always be less than or equal to 90 degrees and will always be referenced from the horizontal or x axis.

To conclude our example we need to determine the magnitude of the resultant force (Figure 1-37). This is accomplished using Pythagorean theorem:

$$R = \sqrt{\Sigma F_y^2 + \Sigma F_x^2}$$

$$= \sqrt{(32.7 \text{ lbs})^2 + (5 \text{ lbs})^2} = 33 \text{ lbs}$$

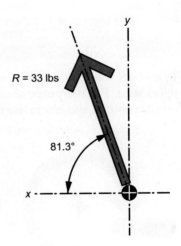

FIGURE 1-37

1.6 RESULTANT OF PARALLEL FORCE SYSTEMS—MOMENTS

To understand how the resultant of a parallel force system is calculated, we must first understand the key concept *moment of force*. In everyday terminology we call this *torque*. For example, torque is the force necessary to loosen or tighten lug bolts on a tire. As we can see in Figure 1-38 on page 20, torque or moment of force is nothing more than a force times the perpendicular distance from the point of rotation to

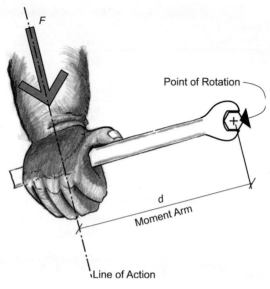

FIGURE 1-38

the line of action of the force. From changing a tire we also know that the longer the wrench, the greater the torque we can apply (with equal or less effort) to the bolts. This distance along the wrench is called the *moment arm*. Since the moment of force is a force times a distance or moment arm, it is measured in units of kip-feet (k-ft), foot-pounds (ft-lbs), or kilonewton-meters (kN-m).

In Figure 1-38, the gorilla is loosening the bolt since the rotation or moment is counterclockwise. For the purpose of this text, we will indicate counterclockwise rotation as positive and clockwise rotation as negative (Figure 1-39).

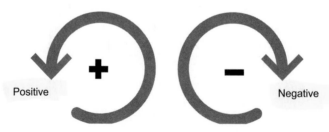

FIGURE 1-39

It is extremely important that we do not confuse the sign convention for forces with the sign convention for moments. When we are summing forces, we use the sign convention for forces (e.g., forces down are negative). When summing moments, we use the sign convention for moments (e.g., a force causing clockwise rotation creates a negative moment).

To establish the resultant of a parallel force system, we need to use moment of force. To demonstrate this, we will introduce a family of gorillas (Figure 1-40).

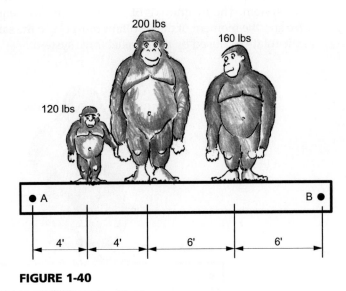

FIGURE 1-40

The first step in determining the resultant force of a parallel force system is to sum up the vertical forces.

$$\Sigma F_y = 120 \text{ lbs} + 200 \text{ lbs} + 160 \text{ lbs} = 480 \text{ lbs}$$

Our gorilla family exerts a total force of 480 pounds, which is the magnitude of the resultant force. Our next step is to take the sum of moments of the individual forces about the point A.

$$\Sigma M_A = -120 \text{ lbs}(4') - 200 \text{ lbs}(8') - 160 \text{ lbs}(14') = -4{,}320 \text{ ft-lbs}$$

Note that this moment is negative since the weights of the gorillas are causing clockwise rotation about the point A. The moment about the point B is positive since the gorillas are causing a counterclockwise rotation about the point B.

The moment created by the gorilla family is equal to the moment of the resultant force. We have already established the magnitude of the resultant force, 480 pounds (the total weight of the gorilla family) and now we must determine how far from the point A to locate the resultant. To accomplish this, we set the moment of the resultant equal to the moment of the family, M_A.

$$M_{\text{Resultant}} = \Sigma M_A$$
$$-480 \text{ lbs}(x) = -4{,}320 \text{ ft lbs}$$
$$x = \frac{-4{,}320 \text{ ft lbs}}{-480 \text{ lbs}} = 9 \text{ ft}$$

Now, if we had a gigantic gorilla weighing 480 pounds standing 9 feet from point A, it would create the same moment as our entire gorilla family in their respective positions (Figure 1-41). The new gorilla constitutes a resultant of a parallel force system. The magnitude of the resultant must equal the sum of the parallel force and the moment of the resultant must cause the same rotation about the point as the rotation caused by the parallel force system.

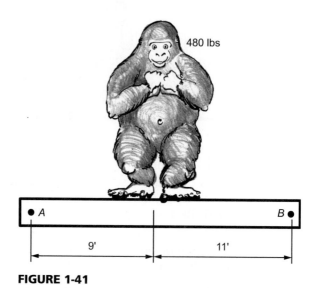

FIGURE 1-41

1.6.1 Moment of a Couple

A *couple* is a moment created by two parallel forces of equal magnitude, with each having an opposite sense of direction. The uniqueness of this moment is that its value remains constant regardless of where the point of rotation occurs. The magnitude of this moment is always F_d where F is the magnitude of one of the forces and d is the distance between them.

In Figure 1-42, we note that the magnitude of the forces is 200 pounds and the distance between them is 6 feet; therefore, the moment created by the couple is

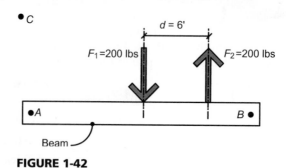

FIGURE 1-42

200 pounds times 6 feet or 1,200 ft-lbs. This moment will remain 1,200 ft-lbs if the point of rotation is A, B, or even point C in Figure 1-42.

1.6.2 Varignon's Theorem

It is often difficult to determine the moment arm of force that is not perpendicular to the point of rotation. Pierre Varignon found a convenient method of calculating the moment of these forces. His theorem states that the moment of a force about a point is equal to the algebraic sum of the moments of the components of the force about the same point.

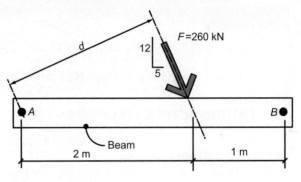

FIGURE 1-43

In Figure 1-43 we see a force of 260 kN on a 12 in 5 slope. It would be cumbersome to determine the moment about the point A or B. It is much easier to break the force into its components as shown in Figure 1-44. To accomplish this, we determine the hypotenuse of the slope, c.

$$c = \sqrt{a^2 + b^2}$$
$$= \sqrt{12^2 + 5^2} = 13$$

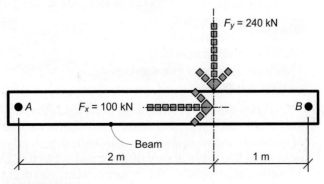

FIGURE 1-44

Now, using proportions shown in Figure 1-44, we can find the magnitude of the components in the x and y direction:

$$\frac{F}{c} = \frac{260 \text{ kN}}{13} = \frac{F_y}{12} = \frac{F_x}{5} = 20 \text{ kN}$$

$$F_y = 12 \, (20 \text{ kN}) = 240 \text{ kN}$$

$$F_x = 5 \, (20 \text{ kN}) = 100 \text{ kN}$$

Having the values of the components, we can take moments about the points A and B. Since the F_x component's line of action extends through both the A and B points, the moment arm, or d distance, is zero and it cannot create moment about either point.

$$M_A = (240 \text{ kN})(2 \text{ m}) = -480 \text{ kN-m}$$

$$M_B = (240 \text{ kN})(1 \text{ m}) = 240 \text{ kN-m}$$

1.7   INTRODUCTION TO LOADS

Up to this point we have dealt with forces acting through a single point, which are referred to *concentrated loads*. There are other loads also supported by the structure. These loads are like pressure in that they are uniformly distributed over an area, and are denoted by "force divided by a unit of area," such as pounds per square inch, kips per square foot, and kilonewtons per square meter. When we think about the loads a structure must support such as snow on a flat roof, we can visualize the meaning of term *loads*. (See Figure 1-45.) Assuming the snow weighs 10 pounds per cubic foot, then we can estimate the load the flat roof supports if it is covered with snow 1'-6" deep. The load must be 1.5 feet times 10 pounds per cubic foot or 15 pounds per square foot. Since the force of the snow load on the roof acts down, it is considered a gravity load. Gravity loads are generally considered static loads, or loads that are not moving. In addition to the static loads, there are dynamic loads, or moving loads. Wind and seismic (earthquake) loads are dynamic loads. We will investigate the effect of these loads on structures later.

In addition to snow loads, there are other gravity loads. These loads are live loads and dead loads. Live loads on floors, for example, are loads caused by people, moveable furniture, and moveable equipment. In the case of roof structures, live loads are rain water and materials and workers on the roof during construction and maintenance. Dead loads include the weight of furniture and equipment and fixed construction materials. In other words, the structure must support its own weight or mass.

When designing structures, dead loads are much easier to estimate than live loads since there are many resources that provide information on standard weights of materials and equipment to assist with dead load estimation. On the other hand, estimating live loads is very difficult and subjective, because live load estimation is a function of probability. For instance, when you design the structure of a classroom or a religious building, how do you determine the number and the weight of the students who will attend a specific class, or parishioners who will attend a religious ceremony? For live loads, we must rely on building codes, which offer live loads based on the appropriate and standard probability of occurrences. For that reason we also rely on the building codes for determining snow loads, which are probabilities derived from climatic studies and vary according to geographic area.

In this chapter, we have studied the forces that must be accounted for when designing a structurally sound building. These forces have weight, direction, and sense of direction, and can be analyzed in their component form. The structure must be able to react to the forces and support the loads from the top of the structure and back to the earth. If the structure does not resist the forces acting on it, the structure will fail. In the next chapter, we will discuss and analyze the forces necessary to keep a building in the state of static equilibrium.

ROOF LOADS

FIGURE 1-45

Sample Problems

Sample Problem 1-1: Determine the components of the force shown in Figure S1-1.

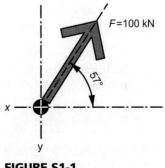

$F=100$ kN

$57°$

x

y

FIGURE S1-1

SOLUTION

Step 1 Investigating Figure S1-1 we notice that the direction is given in terms of an angle with the horizon indicating that we need to use trigonometry to solve for the components. To calculate the horizontal component F_x, we need to use the cosine function.

$$F_x = F(\cos \theta) = 100 \text{ kN}(\cos 57°) = 54.5 \text{ kN}$$

Step 2 To determine the vertical component F_y, we use the sine function (Figure S1-2).

$$F_y = F(\sin \theta) = 100 \text{ kN}(\sin 57°) = 83.9 \text{ kN}$$

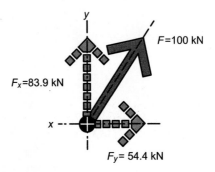

FIGURE S1-2

Sample Problem 1-2: Determine the components of the force shown in Figure S1-3.

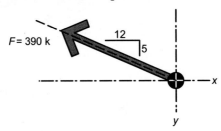

FIGURE S1-3

SOLUTION

Step 1 In this problem we are given a force of 390 kips acting at a slope of 5 in 12 with the horizon. To determine the components of the force, we will determine the hypotenuse of the slope (the c side) and use proportions.

$$c = \sqrt{a^2 + b^2} = \sqrt{5^2 + 12^2} = 13$$

Step 2 To calculate F_x, we establish the following proportion (Figure S1-4):

$$\frac{F_x}{12} = \frac{F}{13}$$

$$F_x = \frac{F(12)}{13} = \frac{390 \text{ k }(12)}{13} = 360 \text{ k}$$

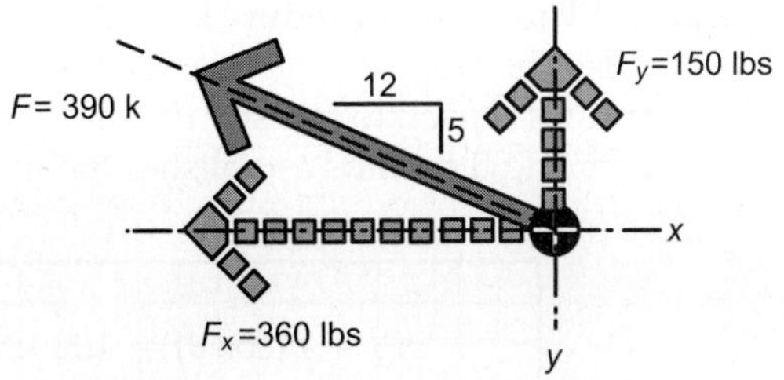

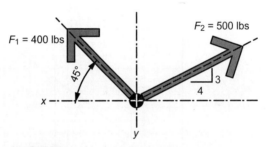

FIGURE S1-4

Step 3 To determine F_y, we use a similar equation:

$$\frac{F_y}{5} = \frac{F}{13}$$

$$F_y = \frac{F(5)}{13} = \frac{390 \text{ k }(5)}{13} = 150 \text{ k}$$

Sample Problem 1-3: Using the parallelogram method, determine the resultant of the force system shown in Figure S1-5.

FIGURE S1-5

SOLUTION

To solve graphic problems, you will need normal drafting equipment including an engineer's scale. Grid paper is helpful when solving graphic as well as computational problems.

Step 1 We begin this problem by accurately redrawing the forces at an appropriate scale (to fit on an $8\frac{1}{2} \times 11$ grid sheet, a scale of 1 inch = 200 pounds is used). Then set the origin at the intersection of a grid and draw the x and y axes. Using a protractor, draw a line at 45 degrees to the horizon. This will become the line of action of F_1. A distance of 2 inches or 400 pounds is then scaled on the line of action from the origin to the northwest, the sense of direction of F_1. (See Figure S1-6.)

The same procedure is used for F_2 except we use the grid paper to establish the slope of 4 in 3. Once the slope is established, a pair of drafting triangles is used to transfer the line of action of F_2 to the origin. The scale of 1 inch = 200 pounds establishes the magnitude of the force.

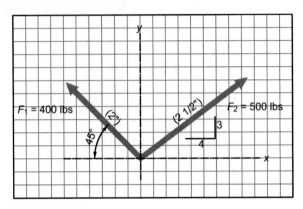

FIGURE S1-6

Step 2   We use a pair of drafting triangles to construct a line parallel to F_2 at the head of F_1. The process is repeated drawing a line parallel to F_1 at the head of F_2. (See Figure S1-7.)

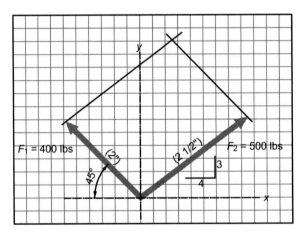

FIGURE S1-7

Step 3   With our parallel lines drawn, we can complete the problem by constructing our resultant by drawing a line from the origin to the intersection of the lines drawn in step 2 (Figure S1-8). The magnitude of the resultant is now scaled to obtain its magnitude and the direction is measured with a protractor for the angle the resultant makes with the horizon. Since the head of the arrow will be at the intersection of the lines and the tail at the origin, the sense of direction will be northeast.

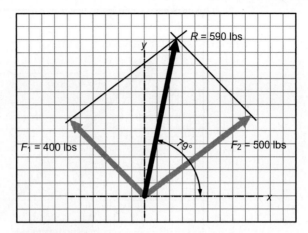

FIGURE S1-8

Sample Problem 1-4: Using the graphic method, determine the resultant of the concurrent force system shown in Figure S1-9.

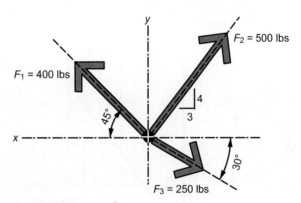

FIGURE S1-9

SOLUTION

Step 1 Using a method similar to the solution of Sample Problem S1-3, we begin by establishing a scale of 1 inch = 200 pounds. Using a protractor we measure the angle of 45° with the horizon and scale the magnitude of 400 pounds.

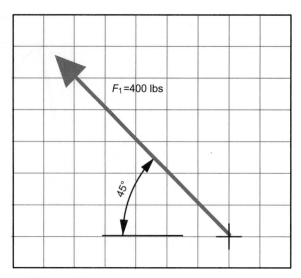

FIGURE S1-10

Step 2 Now that we have the first force drawn, we proceed to draw force F_2. For this force we need to use the grid to acquire the slope of 4 in 3. Using two drafting triangles we transfer the slope to the head of F_1 and scale the magnitude of 500 pounds. All the given forces are drawn head to tail (Figure S1-11).

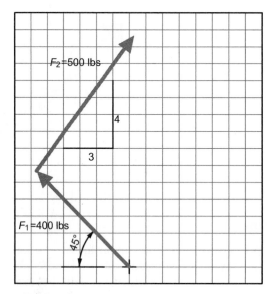

FIGURE S1-11

Step 3 Once again we use our two drafting triangles and the grid to draw a reference line with the head of F_2. Using a protractor we measure an angle of −30 degrees to the horizon (Figure S1-12). F_3 is then drawn to the proper scale.

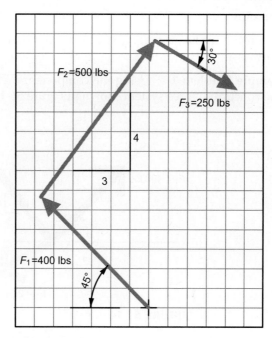

FIGURE S1-12

Step 4 To conclude the problem, we construct the resultant by drawing a line from the tail of F_1 to the head of F_3. We scale the length of the line to obtain the magnitude and using a protractor we are able to establish the direction of the resultant. The sense of direction in this problem is northeast since the resultant is always drawn head to head and tail to tail. (See Figure S1-13.)

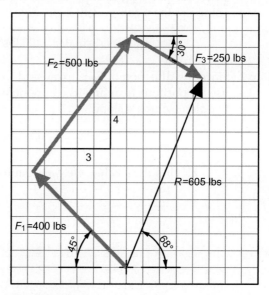

FIGURE S1-13

Results

The resultant has a magnitude of 605 pounds at an angle of 68 degrees to the northeast.

Sample Problem 1-5: Using the component method, determine the resultant of the concurrent force system shown in Figure S1-14.

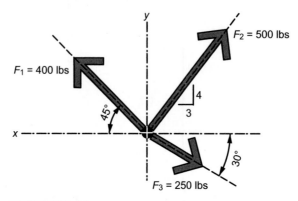

FIGURE S1-14

SOLUTION

Step 1 We begin by calculating the components of all the forces in the system. Since the direction of F_1 is given by an angle, we need to use trigonometry to determine its components.

$$F_{1x} = (F_1) \cos \theta = (400 \text{ lbs}) \cos 45° = -282.8 \text{ lbs}$$

Since the sense F_{1x} is right to left, the x component is negative and F_{1y}, going up, the y component is positive.

$$F_{1y} = (F_1) \sin \theta = (400 \text{ lbs}) \sin 45° = 282.8 \text{ lbs}$$

Step 2 Since the direction of F_2 is expressed in terms of a slope, we will need to use proportions to define its components. First, we have to calculate c, the hypotenuse of the slope:

$$c = \sqrt{a^2 + b^2} = \sqrt{4^2 + 3^2} = 5$$

Having the c value, we can establish our proportions:

$$\frac{F_{2x}}{b} = \frac{F_{2y}}{a} = \frac{F_2}{c}$$

$$\frac{F_{2x}}{3} = \frac{F_{2y}}{4} = \frac{F_2}{5} = \frac{500 \text{ lbs}}{5} = 100 \text{ lbs}$$

$$F_{2x} = 3(100 \text{ lbs}) = 300 \text{ lbs}$$

$$F_{2y} = 4(100 \text{ lbs}) = 400 \text{ lbs}$$

Since the sense of the x component is left to right and the sense of the y component is going up, both are positive.

Step 3   Once again we need to employ trigonometry to resolve the components of F_3.

$$F_{3x} = F \cos \theta = (250 \text{ lbs}) \cos 30° = 216.5 \text{ lbs}$$

$$F_{3y} = F \sin \theta = (250 \text{ lbs}) \sin 30° = -125.0 \text{ lbs}$$

Since the sense of the y component is down, it is negative. The x component, however, is going left to right and therefore must be positive.

Step 4   Having all the components of the given forces, we now summarize the component in the x and y direction.

$$\Sigma F = F_{1y} + F_{2y} + F_{3y} = 282.8 \text{ lbs} + 400 \text{ lbs} - 125.0 \text{ lbs} = 557.8 \text{ lbs}$$

and

$$\Sigma F_x = F_{1x} + F_{2x} + F_{3x} = -282.8 \text{ lbs} + 300 \text{ lbs} + 216.5 \text{ lbs} = 233.7 \text{ lbs}$$

Step 5   Using the Pythagorean theorem, we can now determine the magnitude of the resultant.

$$R = \sqrt{\Sigma F_y^2 + \Sigma F_x^2} = \sqrt{(557.8 \text{ lbs})^2 + (233.7 \text{ lbs})^2} = 604.8 \text{ lbs}$$

Step 6   To determine the direction of the resultant, we use the inverse tangent.

$$\tan^{-1} \theta = \frac{\Sigma F_y}{\Sigma F_x} = \tan^{-1}\left(\frac{557.8 \text{ lbs}}{233.7 \text{ lbs}}\right) = 67.3°$$

Results

Since we can now state that the magnitude of the resultant is 604.8 pounds and since ΣF_y and ΣF_x are both positive, the direction of the resultant is 67.3 degrees to the northeast.

Comparing the results of Sample Problem 1-4 and Sample Problem 1-5, we see a 0.03 % difference in the magnitude of the resultant and a 1.03% difference in the angle with the horizon. The component method is the most accurate of the two methods, but the graphic method does produce acceptable accuracy for structures.

Sample Problem 1-6: Determine the resultant of the parallel force system acting on a beam as shown in Figure S1-15.

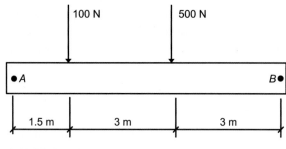

FIGURE S1-15

SOLUTION

Step 1 We begin this problem by determining the magnitude and direction of the resultant. Logically the magnitude of the resultant and direction can be obtained by summing the forces.

$$R = \Sigma F = (-100 \text{ N}) + (-500 \text{ N}) = -600 \text{ N}$$

The resultant is vertical and since the sign of the resultant is negative, the sense is down.

Step 2 Now that we have the magnitude and direction of the resultant, we can find its location on the beam by taking moments. We can take moments about either end of the beam so we will begin by using the point A and verify the answer using point B.

Since we are dealing with moments, we must realize the signs are a result of rotation and not direction.

$$\Sigma M_A = -(100 \text{ N})(1.5 \text{ m}) - (500 \text{ N})(4.5 \text{ m}) = -2{,}400 \text{ Nm}$$

Note that both the moments of the forces and the resultant are negative, which indicates the moments are creating clockwise rotation of the beam about the point A.

If we equate the moment about point A with the moment of the resultant, we can locate the resultant on the beam.

$$\Sigma M_A = M_{Resultant}$$
$$-2{,}400 \text{ Nm} = (-600 \text{ N})d$$
$$d = \frac{-2{,}400 \text{ Nm}}{-600 \text{ N}} = 4 \text{ m (from point } A)$$

Results

The magnitude of the resultant is 600 N vertical force acting down at 4 meters from the point A as shown in Figure S1-16.

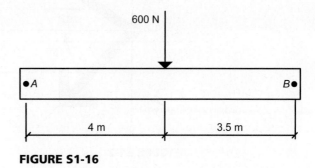

FIGURE S1-16

Check our solution by taking moments about point B.

$$\Sigma M_B = (100 \text{ N})(6 \text{ m}) + (500 \text{ N})(3 \text{ m}) = 2{,}100 \text{ Nm}$$

$$\Sigma M_B = M_{\text{Resultant}}$$

$$2{,}100 \text{ Nm} = (600 \text{ N})(d)$$

$$d = \frac{2{,}100 \text{ Nm}}{600 \text{ m}} = -3.5 \text{ m}$$

Results

These calculations verify that our original determination of the resultant was correct.

Supplementary Exercises

1-1. Determine the components (F_x and F_y) of the forces shown in Figure E1-1.

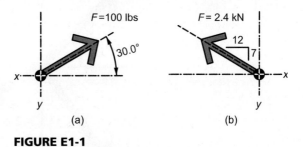

FIGURE E1-1

1-2. Using the parallelogram method, determine the magnitude, direction, and sense of the resultant of the concurrent force system shown in Figure E1-2.

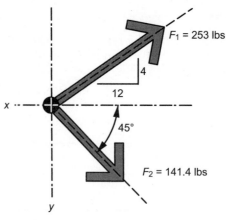

FIGURE E1-2

1-3. Using the parallelogram method, determine the magnitude, direction, and sense of the resultant of the concurrent force system shown in Figure E1-3.

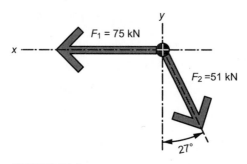

FIGURE E1-3

1-4. Determine the magnitude, direction, and sense of the resultant of the concurrent force system shown in Figure E1-4 using the graphic method.

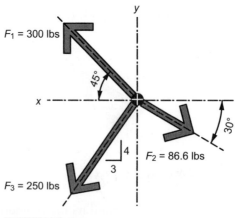

FIGURE E1-4

1-5. Determine the magnitude, direction, and sense of the resultant of the concurrent force system shown in Figure E1-5 using the graphic method.

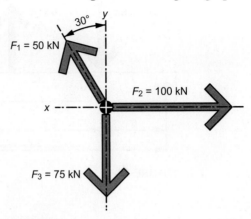

FIGURE E1-5

1-6. Verify your answer to Exercise 1-2 using the component method of determining the resultant of a concurrent force system.

1-7. Using the component method, verify your answer to Exercise 1-4.

1-8. Using the component method, verify your answer to Exercise 1-5.

1-9. Determine the resultant of the parallel force system shown in Figure E1-6.

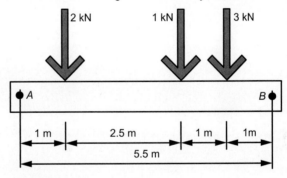

FIGURE E1-6

1-10. Determine the moments of the forces about the point A in Figure E1-7.

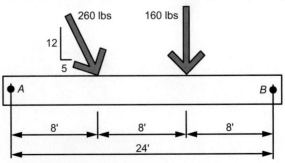

FIGURE E1-7

1-11. Determine the moments of the forces about the point B in Figure E1-8.

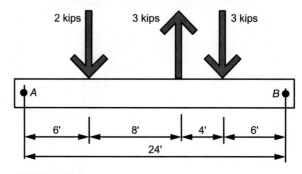

FIGURE E1-8

CHAPTER 2

Static Equilibrium and Reactions

As Newton (Figure 2-1) observed, a building will remain at rest (the state of static equilibrium) when all the forces acting on it (applied loads) are resisted by equal and opposite reactions.

For every action there is an equal and opposite reaction

Sir Isaac Newton

FIGURE 2-1

2.1 STATIC EQUILIBRIUM

Figuratively speaking, we do not want our building to move (even though we know that all buildings expand and contract and that the top of skyscrapers sway in the wind). To prevent them from moving in ways we do not intend, we must keep them in *static equilibrium.* A structure is in static equilibrium when the sum of the forces in the x direction is equal to zero, the sum of the forces in the y direction is equal to zero, and the sum of the moments is equal to zero. Therefore, we have three equations to satisfy to ensure our building's stability:

$$\Sigma F_x = 0$$

$$\Sigma F_y = 0$$

$$\Sigma M = 0$$

If we are able to analyze a structure with these three equations, we call the structure *determinate*. If more than three equations are required to analyze the structure, it is *indeterminate* and beyond the scope of this text.

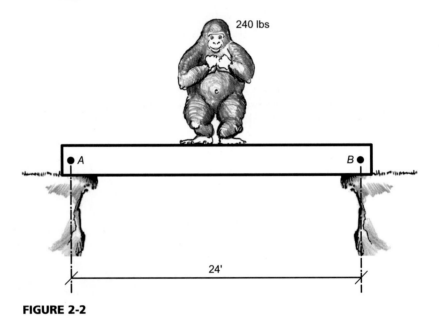

FIGURE 2-2

For example, if a gorilla stands on a beam, we know that the beam has to be strong enough to resist the gorilla's weight and transfer it to the ground. In Figure 2-2 above, the gorilla weighs 240 pounds and we assume the beam is in static equilibrium since it is at rest. For the beam to remain at rest, there must be some forces at the end supports acting up to oppose the gorilla's weight. These forces are called *reactions*.

2.2 REACTIONS

Since the weight of the gorilla (an applied load) is acting at the center of the beam, we can logically conclude that each support must supply a reaction of 120 pounds or one-half the gorilla's weight, if the beam is in static equilibrium. The beam can now be drawn as shown in Figure 2-3. Reactions are indicated as forces with a short diagonal line drawn through the center.

Since there is no force in the x direction, we can assume there are no horizontal reactions. Now that we have balanced the forces, we can test the beam to determine if the moments are equal to zero. We will begin by taking moments about the A support using our previously established standard that clockwise moments are negative and counterclockwise moments are positive.

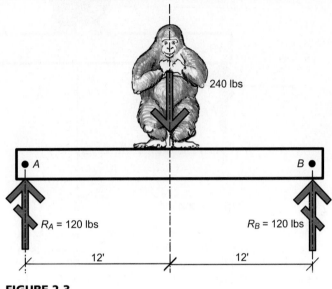

FIGURE 2-3

$$\Sigma M_A = 0$$
$$= -240 \text{ lbs } (12 \text{ ft}) + 120 \text{ lbs } (24 \text{ ft}) = 0$$

Since the gorilla is standing in the mid-span and the beam is symmetrically loaded, we would obtain the same results if we were to take the moments about the support B. But what would be the result if we would calculate the moments about the point C, 4 feet to the right of support A (Figure 2-4)? If the beam is in *static equilibrium*, we should be able to take moments at any point and still have the result equal to zero.

$$\Sigma M_c = 0$$
$$= -120 \text{ lbs } (4 \text{ ft}) - 240 \text{ lbs } (8 \text{ ft}) + 120 \text{ lbs } (20 \text{ ft}) = 0$$

We can, therefore, safely conclude that the beam with the gorilla standing at mid-span is in *static equilibrium*.

2.3   FREE BODY DIAGRAMS

We now replace the supports in the previous example with two gorillas (Figure 2-5). Assume that each of the supporting gorillas must supply upward force of 120 pounds at the points A and B. We have created a structural system similar to a

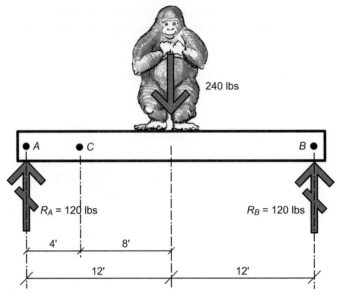

FIGURE 2-4

building girder supported by columns. This newly created system demonstrates two very important points: (1) if the structural system is in *static equilibrium*, then all of the components of the system are in *static equilibrium*, and (2) as mentioned in Chapter 1, the structural system conveys all the loads or forces back to the earth.

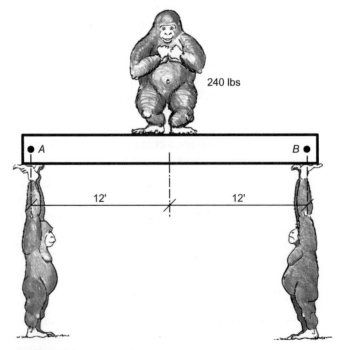

FIGURE 2-5

If our structural system is in *static equilibrium*, then each of our supporting gorillas is in static equilibrium. To explore this issue, we will isolate the gorilla at support *B* (Figure 2-6).

B

Reaction
Force

Applied Force
of the Beam

Reaction
Force

FIGURE 2-6

The end of the beam requires a reaction of 120 pounds, which is transferred to the gorilla as a 120-pound force (Figure 2-6). To keep the gorilla in static equilibrium, the sum of the forces must be equal to zero; therefore, the earth must supply a reaction of 120 pounds. Since all of the forces in Figure 2-6 have the same line of action, there are no moments acting on the gorilla.

Often we get involved in formulas and fail to rely on common sense. So, before we investigate equilibrium equations in resolving structural problems, we will use simple proportions to understand the relationship between applied loads and reactions. What if we move the 240-pound gorilla from mid-span to a position 16 feet to the right of support *A*? (See Figure 2-7 on page 44.)

Using logic we can safely assume that the gorilla at support *B* must assume a larger share of the load or weight of the 240-pound gorilla. The gorilla moved 4 feet closer to the *B* support. In fact, the gorilla at support *B* must support 4/24 or 1/6 more load or:

$$R_B = 120 \text{ lbs} + 1/6 \ (240 \text{ lbs}) = 160 \text{ lbs}$$

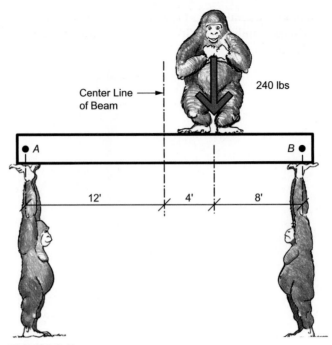

FIGURE 2-7

Conversely, the gorilla at support *A* has a lighter load since it is reduced by 1/6.

$$R_A = 120 \text{ lbs} - 1/6 \,(240 \text{ lbs}) = 80 \text{ lbs}$$

Suppose we move the gorilla once again, this time to a position 18 feet to the right of support *A* (Figure 2-8)?

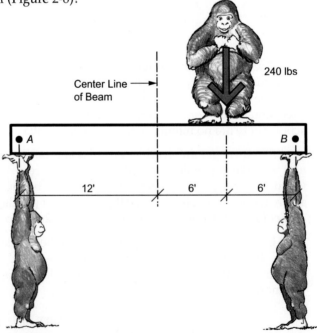

FIGURE 2-8

The gorilla has now moved 6 feet from the center of the beam toward support B indicating that the gorilla at B must support 6/24 or 1/4 more load than when the 240-pound gorilla was located at the center. In this case:

$$R_B = 120 \text{ lbs} + 1/4\,(240) \text{ lbs} = 180 \text{ lbs}$$

and

$$R_A = 120 \text{ lbs} - 1/4\,(240 \text{ lbs}) = 60 \text{ lbs}$$

Reviewing these examples, we can draw a few simple conclusions: (1) the closer the applied loads move towards a support, the larger the reaction becomes at that support and (2) the magnitude of the increased force that the reaction must support is directly proportional to the distance that the applied load is moved in the direction of that support. If we keep these conclusions in mind when dealing with structural problems, we will be less apt to make mistakes.

Now we will return to the use of the equilibrium equations and see if we obtain the same results as we did using proportions. We begin by drawing Figure 2-7, using structural symbols.

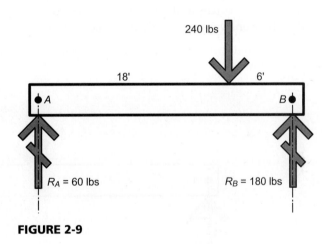

FIGURE 2-9

Assume the beam is in static equilibrium and take moments about the point A. As always, we will use our previously established standard that clockwise moments are negative and counterclockwise moments are positive.

$$\Sigma M_A = 0$$
$$0 = -240 \text{ lbs } (18 \text{ ft}) + R_B(24 \text{ ft})$$
$$-R_B = \frac{-4{,}320 \text{ ft} - \text{lbs}}{24 \text{ ft}}$$
$$R_B = 180 \text{ lbs}$$

Now that we have one of the reactions, we can use the sum of forces in the y direction to determine the reaction at A.

$$\Sigma F_Y = 0$$
$$0 = -240 \text{ lbs} + 180 \text{ lbs} + R_A$$
$$-R_A = -60 \text{ lbs}$$
$$R_A = 60 \text{ lbs}$$

Since there are no horizontal forces acting on the beam, we know the sum of forces in the x direction is equal to zero.

We can check our answer by using the sum of moments about B.

$$\Sigma M_B = 0$$
$$0 = 240 \text{ lbs (6 ft)} - R_A \text{ (24 ft)}$$
$$R_A = \frac{1,440 \text{ ft} - \text{lbs}}{24 \text{ ft}} = 60 \text{ lbs} \quad \text{OK}$$

By taking moments about the point B, we obtained the same reaction as we did when we summed forces in the y direction; therefore, our calculations are correct. Notice that these answers are exactly the same as those obtained using proportions.

We will conclude this section by using the equations to verify our answers for Figure 2-10.

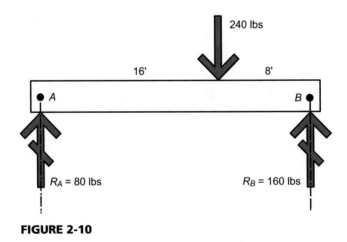

FIGURE 2-10

$$\Sigma M_A = 0 = -240 \text{ lbs (16 ft)} + R_B \text{ (24 ft)}$$
$$-R_B = \frac{-3,840 \text{ ft} - \text{lbs}}{24 \text{ ft}}$$

$$R_B = 160 \text{ lbs}$$
$$\Sigma F_Y = 0$$
$$0 = -240 \text{ lbs} + 160 \text{ lbs} + R_A$$
$$R_A = 80 \text{ lbs}$$

We can check our answer by taking moments about the opposite end, support B.

$$\Sigma M_B = 0 = 240 \text{ lbs } (8 \text{ ft}) - R_A (24 \text{ ft})$$
$$R_A = \frac{1,920 \text{ ft} - \text{lbs}}{24 \text{ ft}} = 80 \text{ lbs} \quad \text{OK}$$

2.4 CONNECTIONS

To further our understanding of reactions, we must examine a few basic types of connections (or joints). These connections are classified according to their ability for the connected member to translate (move in one direction) or rotate. The categories of connections are *roller, pin,* and *fixed.*

2.4.1 The Roller

The term *roller* refers to connections that are able to transfer load perpendicular to the surface they rest on but allow the connected member to rotate and move, or translate parallel to the surface it rest on. These types of connections are usually used on the exterior structures where large ranges in temperature occur, such as bridges. Even though these connections are seldom encountered in buildings, we use them in elementary structural analysis, since the roller, when combined with a pinned connection, provides us with a simple supported beam, a determinate beam that can be analyzed using static equilibrium.

When a structural member, such as a beam, is supported on one end by a roller, there is only one reaction (Figure 2-11). If the surface is horizontal, the opposing reaction is vertical. With the roller connection the member is free to rotate. Figure 2-11 on page 48 shows its structural symbol.

2.4.2 The Pinned Connection

In building construction the pinned connection is the most common connection. It is able to transfer load from the connected member in any direction, yet it allows the member to rotate. The majority of wood and steel beam to girder connections and girders to column connections are pinned connections.

If we remove the roller and replace it with a pinned connection (Figure 2-12), we would still have only one reaction. In this case, we have no idea of its direction. To determine the magnitude and direction of the reaction, we will have to depend on the x and y components of the reaction. This type of connection allows the connected member to rotate. Figure 2-12 on page 48 shows its structural symbol.

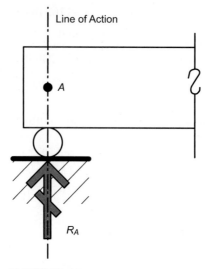

FIGURE 2-11

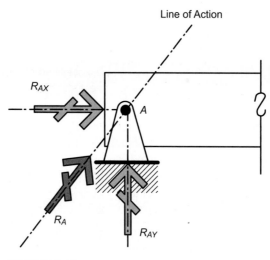

FIGURE 2-12

2.4.3 Fixed Connection

The fixed connection transfers loads from the connected member in any direction but does not allow the member to rotate. Since it prevents rotation, this type of joint must supply a moment reaction when the member tries to rotate under applied load. This type of connection is common in poured-in-place concrete structures and in many welded steel structures. The cantilevered beam (which will be covered later in this chapter) is an excellent example of a fixed connection.

If we want to have a connection that prevents the connected member from rotating, we would use a fixed connection (Figure 2-13). Like the pinned connection, the fixed connection has one reaction. The magnitude and direction of the reaction is unknown, and we must determine the components of the reaction to obtain these values. Since the fixed connection prevents the connected member from rotating, we also have a reaction moment.

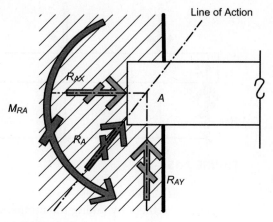

FIGURE 2-13

2.5 ANALYZING BEAM REACTIONS

Now that we have covered the basics of static equilibrium, we can examine its application. In Figure 2-14, we have a simple beam: 18 feet long with a roller at support A and a pin on the right at support B. The beam supports two applied loads: a 1.5 kip force at 3 feet to the right of A and a 3.0 kip force at mid-span. Under this loading condition, what are the beam reactions?

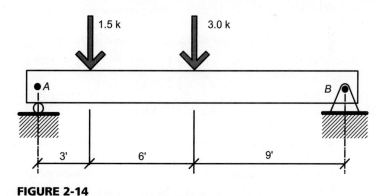

FIGURE 2-14

Always start a solution by examining the problem and determining what data are given. In this problem we have two connections: a roller at support A that has one reaction perpendicular to the surface it rests on; and a pinned connection that has two unknowns, the x and y components of the reaction B (Figure 2-15). Since there are no horizontal forces, we can assume that the sum of forces in the x

direction is equal to zero. Relying on our previous exercises, we know that a larger portion of the loads will be supported at A. Since there are no horizontal forces, we can begin by taking the moments about either support A or B.

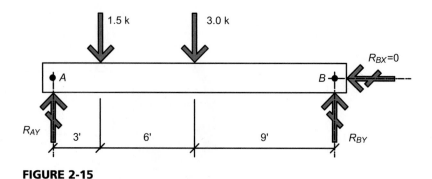

FIGURE 2-15

We will begin the problem by taking the moments about support A.

$$\Sigma M_A = 0$$
$$0 = -1.5 \text{ kips (3 ft)} - 3 \text{ kips (9 ft)} + R_{BY} \text{ (18 ft)}$$
$$-R_{BY} = \frac{-4.5 \text{ kip-ft} - 27 \text{ kip-ft}}{18 \text{ ft}}$$
$$R_{BY} = 1.75 \text{ kips}$$

Now that we have one of the reactions, we can solve for R_A by summing the vertical forces.

$$\Sigma F_Y = 0$$
$$0 = -1.5 \text{ kips} - 3 \text{ kips} + 1.75 \text{ kip} + R_A$$
$$-R_A = -2.75 \text{ kips}$$
$$R_A = 2.75 \text{ kips}$$

To check our answer, we will take the moments about the B support.

$$\Sigma M_B = 0$$
$$0 = 3 \text{ kips (9 ft)} + 1.5 \text{ kips (15 ft)} - R_A \text{ (18 ft)}$$
$$R_A = \frac{27 \text{ kip-ft} + 22.5 \text{ kip-ft}}{18 \text{ ft}} = 2.75 \text{ kips} \quad \text{OK}$$

In this next example, we will look at a beam with a fixed end, a cantilever (Figure 2-16). We begin by examining the problem to determine the given data. Support A is a fixed connection that must supply an x and y component of the

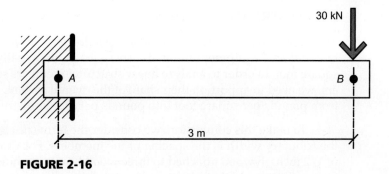

FIGURE 2-16

reaction force along with a moment reaction, M_{RA}. The B end of the beam has no support and therefore is termed a *free* end. Since there are no horizontal applied forces acting on the beam, there is no horizontal or x component of the reaction. The only reaction force is the vertical or y component, which must equal the applied load at B (Figure 2-17).

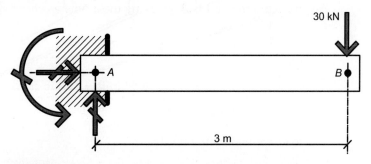

FIGURE 2-17

To obtain a value for the reaction moment, M_{RA}, take the moments about A.

$$\Sigma M_A = 0$$

$$0 = -30 \text{ kN } (3 \text{ m}) + M_{RA}$$

$$-M_{RA} = -90 \text{ kN-m}$$

$$M_{RA} = 90 \text{ kN-m}$$

The only reaction force is the vertical or y component, which must equal the applied load at B.

$$\Sigma F_y = 0$$

$$0 = -30 \text{ kN } + R_{Ay}$$

$$-R_{Ay} = -30 \text{ kN}$$

$$R_{Ay} = 30 \text{ kN}$$

2.6 DISTRIBUTED LOADS

The loads a structure must support are usually derived from codes or provided by material and equipment manuals. These loads are usually given in pounds per square foot. In order to analyze linear structural members such as beams and girders, we need to apportion their share of the overall system, converting these loads from pounds per square foot into pounds per foot or distributed loads (w).

To make this conversion, we consider the term *tributary width*. In most cases the tributary width is the spacing of the members. For example, we have a sheet of 1/2 inch plywood attached to three wood beams as shown in Figure 2.18. Nine inches of sand, weighing 100 pounds per cubic foot, is then placed on the plywood.

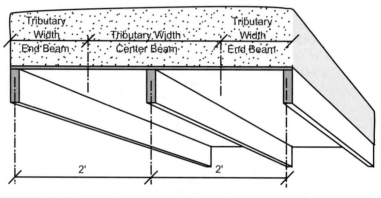

FIGURE 2-18

The load applied by the sand is:

$$\text{Weight of sand} = \frac{9 \text{ in}}{12 \text{ in/ft}}\left(\frac{100 \text{ lbs}}{\text{ft}^3}\right) = \frac{75 \text{ lbs}}{\text{ft}^2}$$

According to the American Plywood Association, 1/2 inch thick plywood weighs 1.5 lbs/ft²; therefore, the *total load* (T) the beams must support is the sum of the weight of the sand and the plywood.

$$T = \frac{75 \text{ lbs}}{\text{ft}^2} + \frac{1.5 \text{ lbs}}{\text{ft}^2} = \frac{76.5 \text{ lbs}}{\text{ft}^2}$$

Since the spacing of the beams supporting the plywood is 2 feet on center, the tributary width for the center beam is 2 feet. The beam supports one foot of sand on each side. To change this total weight or total load (T) into a distributed load, we multiply the total load by the tributary width.

$$\omega = T \times \text{trib. width}$$

$$= \frac{76.5 \text{ lbs}}{\text{ft}^2}(2 \text{ ft}) = 153 \text{ lbs/ft}$$

Structurally we can represent the beam as shown in Figure 2-19. If we examine the end beams that support the edge of the plywood (Figure 2-18), we see that

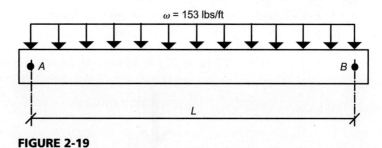

FIGURE 2-19

their tributary width is only one foot, plus half the beam width. Since the edge beams support only half of the load that the center beam supports, ω would be approximately 76.5 pounds per foot for these beams.

The type of load shown in Figure 2.19 is termed *uniform distributed load*, since it is of uniform magnitude. To calculate the reactions for a beam supporting this type of load, we can readily convert the uniform distributed load into a concentrated load by multiplying ω by the length of the load, which is denoted by L_L, to distinguish it from the length of the beam, L. A concentrated load, which has been converted from the distributed load, is designated by W, where $W = \omega L_L$.

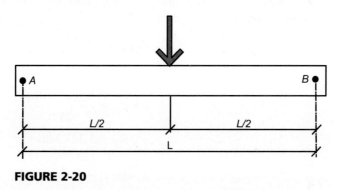

FIGURE 2-20

The location of W is always at the center of the loaded area (Figure 2-20). If the uniform distributed load covered only 16 feet of a 24-foot-long beam, as shown in Figure 2-21, W would be located at 8 feet from A.

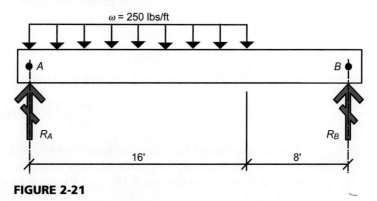

FIGURE 2-21

In this example, we convert the uniform load into a concentrated load acting at the center of the 16 feet of distributed load (Figure 2-22).

$$W = \omega L_L = 250 \text{ lbs (16 ft)} = 4{,}000 \text{ lbs}$$

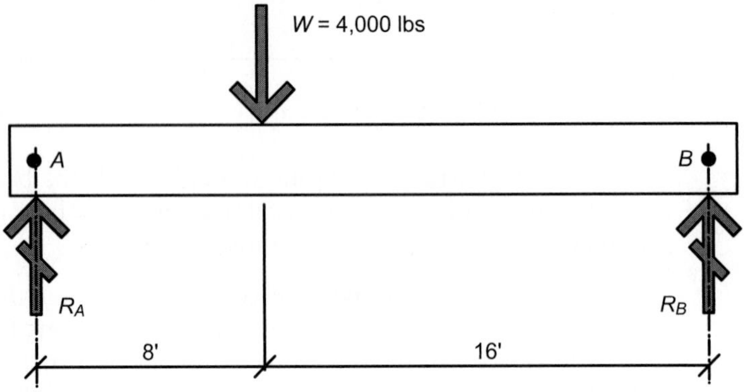

FIGURE 2-22

Having converted the uniform distributed load into a concentrated load, we continue by resolving the reactions.

$$\Sigma M_A = 0$$

$$0 = -4{,}000 \text{ lbs (8 ft)} + R_B(24 \text{ ft})$$

$$-R_B = \frac{32{,}000 \text{ ft-lbs}}{24 \text{ ft}}$$

$$R_B = 1{,}333.3 \text{ lbs}$$

$$\Sigma F_Y = 0$$

$$0 = -4{,}000 \text{ lbs} + 1{,}333.3 \text{ lbs} + R_A$$

$$-R_A = -2{,}666.7 \text{ lbs}$$

$$R_A = 2{,}666.7 \text{ lbs}$$

Check:

$$\Sigma M_B = 0$$

$$0 = 4{,}000 \text{ lbs (16 ft)} - R_A (24 \text{ ft})$$

$$R_A = \frac{64{,}000 \text{ ft-lbs}}{24 \text{ ft}} = 2{,}666.7 \text{ lbs} \quad \text{OK}$$

Not all distributed loads are uniform. For example, there are linear distributed loads. If you imagine the profile of snow drift against a wall or sand shifted on the piece of plywood (Figure 2-23), you have a triangular load. In the case of the sand on the plywood, if we assume the sand is 9 inches deep at the high point, the load would vary from 76.5 pounds per square foot and diminishes to 0 pounds per square foot at the opposite end (Figure 2-23).

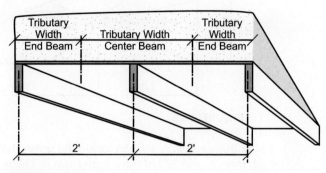

FIGURE 2-23

Using the tributary width of the center beam, the load on the high end of the plywood is:

$$\omega = T \times \text{trib. width} = \frac{76.5 \text{ lbs}}{\text{ft}^2} (2 \text{ ft}) = 153 \text{ lb/ft}$$

Structurally the beam can be drawn along its length (Figure 2-24).

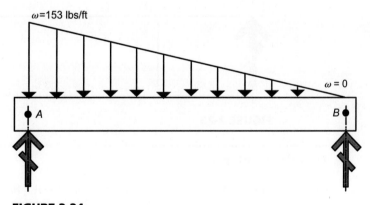

FIGURE 2-24

To convert the linear distributed load into a concentrated load, we determine where the center of its area (termed *centroid* in structures) is located. From geometry we know that drawing lines from the mid-point of each side to the vertex of the opposite angle will give us the center of the area at the intersection of these lines (Figure 2-25, page 56).

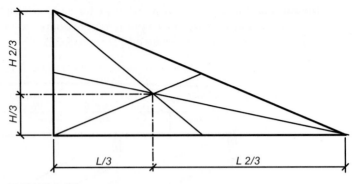

FIGURE 2-25

Using this geometry we can locate the converted concentrated load at 1/3 of the length from A and 2/3 of the length from B. The magnitude of the concentrated load will be determined by its area.

$$W = 1/2\ \omega L_L$$

The beam may now be drawn as shown in Figure 2-26.

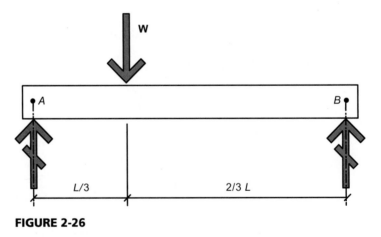

FIGURE 2-26

Sample Problems

Sample Problem 2-1: Determine the reactions for the beam in Figure S2-1.

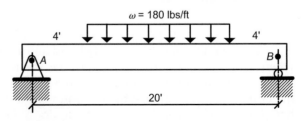

FIGURE S2-1

SOLUTION

Step 1 In Figure S2-1, we find a uniform distributed load on a beam with a pinned connection at support A and a roller at support B. The uniform distributed load is vertical; therefore, there is no horizontal component of the reaction at A.

Step 2 Our first challenge is to determine the resultant of the uniform distributed load, W. The length of the load on the beam, L_L, will be the total length minus the distance from support A to the load and the distance from the load to support B.

$$L_L = 20 \text{ ft} - 4 \text{ ft} - 4 \text{ ft} = 12 \text{ ft}$$

$$W = \omega L_L = 180 \text{ lbs/ft} (12 \text{ ft}) = 2,160 \text{ lbs}$$

W will be located at the center of L_L or 10 feet to the right of support A.

Step 3 Having calculated L_L and W, we can redraw the beam as shown in Figure S2-2. Since the resultant, W, acts at the center of L_L, we realize that W is located at the center of the 20-foot span; therefore, the reactions must be equal or half of W.

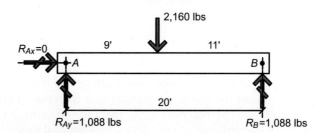

FIGURE S2-2

$$R_A = R_B = \frac{W}{2} = \frac{2,160 \text{ lbs}}{2} = 1,080 \text{ lbs}$$

We can verify our logical conclusion by taking moment about the support A and summing forces.

$$\Sigma M_A = 0$$

$$0 = -2,160 \text{ lbs} (10 \text{ ft}) + R_B(20 \text{ ft})$$

$$-R_B = \frac{-2,160 \text{ lbs} (10 \text{ ft})}{20 \text{ ft}} = -1,080 \text{ lbs}$$

$$R_B = 1,080 \text{ lbs}$$

$$\Sigma F_y = 0$$

$$0 = -12,060 \text{ lbs} + 1,080 \text{ lbs} + R_A$$

$$-R_A = -1,080 \text{ lbs}$$

$$R_A = 1,080 \text{ lbs}$$

Results

The calculations verify our initial conclusion that the reactions are equal.

Sample Problem 2-2: Calculate the reactions for the beam with an overhang or cantilevered end shown in Figure S2-3.

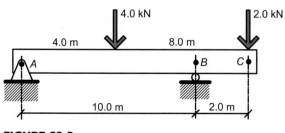

FIGURE S2-3

SOLUTION

Step 1 Once again, when we examine the loads on the beam in S2-3, we see that all the loads are vertical and therefore there is no horizontal component of the reaction at A, the pinned joint. Even though the beam has an overhang, we can follow our normal procedure for determining reactions. (See Figure S2-4.)

$$\Sigma M_A = 0$$

$$0 = -4 \text{ kN } (4 \text{ m}) - 2.0 \text{ kN } (10 \text{ m}) + R_B (8 \text{ m})$$

$$-R_B = \frac{-16 \text{ kNm} - 20 \text{ kNm}}{8 \text{ m}} = -4.5 \text{ kN}$$

$$R_B = 4.5 \text{ kN}$$

$$\Sigma F_y = 0$$

$$0 = -4.0 \text{ kN} - 2.0 \text{ kN} + 4.5 \text{ kN} + R_A$$

$$R_A = 1.5 \text{ kN}$$

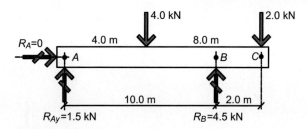

FIGURE S2-4

Step 2 If we wish to verify our answer, we can do so by taking moments about the B support.

$$\Sigma M_B = 0$$

$$0 = -2.0 \text{ kN} (2.0 \text{ m}) + 4.0 \text{ kN} (6.0 \text{ m} - 2.0 \text{ m}) - R_A(8.0 \text{ m})$$

$$R_A = \frac{-4.0 \text{ kNm} + 16.0 \text{ kNm}}{8.0 \text{ m}} = 1.5 \text{ kN}$$

Results

Taking moments about the support B verifies that our original calculations are correct.

Sample Problem 2-3: Calculate the reactions for the beam shown in Figure S2-5.

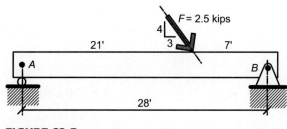

FIGURE S2-5

SOLUTION

Step 1 This problem is an example of how the uses of components of a force make the analysis of beams more convenient when using trigonometry to resolve the moments about A. To analyze this beam, we begin by breaking the 2.5 kip force into its x and y components. (See Figure S2-6.)

$$c = \sqrt{a^2 + b^2} = \sqrt{4^2 + 3^2} = 5$$

$$\frac{3}{5} = \frac{F_x}{F}$$

$$F_x = \frac{3(2.5 \text{ kips})}{5} = 1.5 \text{ kips}$$

$$\frac{4}{5} = \frac{F_y}{F}$$

$$F_y = \frac{4(2.5 \text{ kips})}{5} = 2.0 \text{ kips}$$

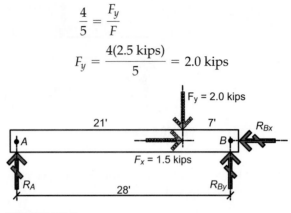

FIGURE S2-6

Step 2   Now that we have the components of the force F, we can resolve the reactions. We will begin by summing the forces in the x direction.

$$\Sigma F_x = 0$$

$$0 = -1.5 \text{ kips} + R_{Bx}$$

$$-R_{Bx} = -1.5 \text{ kips}$$

$$R_{Bx} = 1.5 \text{ kips}$$

Step 3   Having resolved the forces in the x direction, we will proceed to calculate the forces in the y direction by taking moments about support B and summing forces in the y direction.

$$\Sigma M_B = 0$$

$$0 = 2.0 \text{ kips (7 ft)} - R_A(21 \text{ ft})$$

$$R_A = \frac{14 \text{ kip-ft}}{21 \text{ ft}} = .66 \text{ kips}$$

$$\Sigma F_y = 0$$

$$0 = -2.0 \text{ kips} + .66 \text{ kips} + R_{By}$$

$$-R_{By} = -1.33 \text{ kips}$$

$$R_{By} = 1.33 \text{ kips}$$

Note

When taking moments about the support B, we realize that the line of action of the horizontal component of the force and the reaction extend through the point B and therefore, cannot cause moment about the support.

Step 4   In practice we would rarely be concerned with the actual magnitude or direction of the reaction at support B, since we would, for convenience, resolve all forces, acting on a structure into the x and y direction. Solely for academic reasons, we will determine these values to enforce our understanding of forces and reactions.

$$R_B = \mathbf{2}\ \overline{R_{By}^2 + R_{Bx}^2}$$

$$= \mathbf{2}\ \overline{(1.33\ \text{kips})^2 + (-1.5\ \text{kips})^2} = 2\ \text{kips}$$

$$u = \tan^{-1}\!\left(\frac{R_{By}}{R_{Bx}}\right) = \tan^{-1}\!\left(\frac{1.33\ \text{kips}}{1.5\ \text{kips}}\right) = 55.8°$$

Results

The reaction at support B makes an angle of 55.8 degrees with the horizon. Since R_{Bx} is negative and R_{By} is positive, we know that the sense of direction of the reaction is to the northwest.

Sample Problem 2-4: Determine the reactions at supports A and B in Figure S2-7.

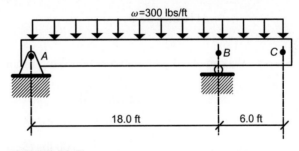

FIGURE S2-7

SOLUTION

Step 1 In this problem we must resolve the uniform distributed load into a concentrated load or resultant force acting in the center of the length of the load.

$$W = \omega L_L = 300\ \text{lbs}\ (18.0\ \text{ft} + 6.0\ \text{ft}) = 7,200\ \text{lbs}$$

Since the length of the load, L_L, is the total length of the beam or 24 feet, the resultant should be located at 12 feet from the A support.

Step 2 The uniform distributed load is vertical as is the resultant of the uniform load; therefore, the horizontal reaction at support A is zero.

Step 3 To resolve the reactions, we will begin by taking moments about the support A and summing forces in the y direction. (See Figure S2-8.)

$$\Sigma M_A = 0$$

$$0 = -7,200\ \text{lbs}\ (12\ \text{ft}) + R_B\ (18\ \text{ft})$$

$$-R_B = \frac{-7,200\ \text{lbs}\ (12\ \text{ft})}{18\ \text{ft}}$$

$$R_B = 4{,}800 \text{ lbs}$$

$$\Sigma F_y = 0$$

$$0 = -7{,}200 \text{ lbs} + 4{,}800 \text{ lbs} + R_A$$

$$-R_A = 2{,}400 \text{ lbs}$$

$$R_A = 2{,}400 \text{ lbs}$$

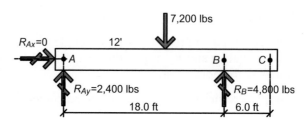

FIGURE S2-8

Step 4 We can check our answers by taking moments about the support B.

$$\Sigma M_B = 0$$

$$0 = 7{,}200 \text{ lbs } (6 \text{ ft}) - R_A(18 \text{ ft})$$

$$R_A = \frac{7{,}200 \text{ lbs } (6 \text{ ft})}{18 \text{ ft}} = 2{,}400 \text{ lbs}$$

Results

The moments about the support B verify our initial calculation of the reactions.

Sample Problem 2-5: Calculate the reactions for the loads shown on the beam in Figure S2-9.

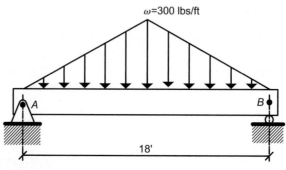

FIGURE S2-9

SOLUTION

Step 1 In this problem we have two linear distributed loads: the first extends from 0 lbs/ft to 300 lbs/ft and the second from 300 lbs/ft to 0 lbs/ft. Our first step involves resolving these loads into resultant concentrated loads and determining their location.

$$W = 1/2\,\omega L_L$$

$$= 1/2\,(300\text{ lbs/ft})\,(9\text{ ft}) = 1{,}350\text{ lbs}$$

The resultant loads are located 2/3 of L_L from the support or 6 feet to the right of support A and 6 feet to the left of support B.

Step 2 The horizontal component of the reaction at support B is zero, since all the loads are vertical. Since the resultant loads are of equal magnitude and symmetrically located, we can logically assume they are equal.

$$R_A = R_B = 1{,}350\text{ lbs}$$

Step 3 We can verify our logical conclusion by taking moments about support A. (See Figure S2-10.)

$$\Sigma M_A = 0$$

$$0 = -1{,}350\text{ lbs (6 ft)} - 1{,}350\text{ lbs (12 ft)} + R_B\text{ (18 ft)}$$

$$-R_B = \frac{-1{,}350\text{ lbs (6 ft)} - 1{,}350\text{ lbs (12 ft)}}{18\text{ ft}}$$

$$R_B = 1{,}350\text{ lbs}$$

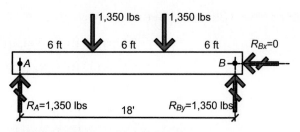

FIGURE S2-10

Results

The reaction calculated by taking moments about support A and summarizing the forces in the y direction verifies our initial conclusion.

Sample Problem 2-6: Determine the reactions for the truss shown in Figure S2-11.

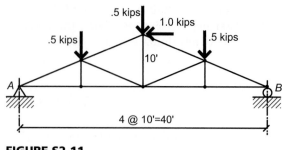

FIGURE S2-11

SOLUTION

We realize immediately that we have a pinned joint at A and a roller at B. Since we have horizontal loads, we can begin by summing the force in the x direction. Taking moments about either support will then supply us with the vertical reaction at B and vertical component at A.

Step 1 The only support that can supply a horizontal component of a reaction is A.

$$\Sigma F_x = 0$$
$$0 = -.25 \text{ kip} + R_{Ax}$$
$$-R_{Ax} = -.25 \text{ kips}$$
$$R_{Ax} = .25 \text{ kips}$$

Step 2 Now that we have the horizontal components, we can resolve the vertical reaction at B and the vertical component at A. (See Figure S2-12.)

$$\Sigma M_A = 0$$
$$0 = -.5 \text{ kips (10 ft)} - .5 \text{ kips (20 ft)} + 1.0 \text{ kips (10 ft)}$$
$$-.5 \text{ kips (30 ft)} + R_B \text{ (40 ft)}$$
$$-R_B = \frac{-20 \text{ kip-ft}}{40 \text{ ft}}$$
$$R_B = .5 \text{ kips}$$

$$\Sigma F_y = 0$$
$$0 = -.5 \text{ kips} - .5 \text{ kips} - .5 \text{ kips} - .5 \text{ kips} + R_{Ay}$$
$$-R_{Ay} = -1.0 \text{ kips}$$
$$R_{Ay} = 1.0 \text{ kips}$$

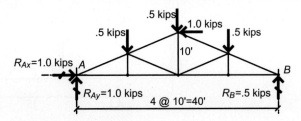

FIGURE S2-12

Results

From this example, we see that analyzing the reactions for a truss is the same as analyzing the reactions for a beam. The moments are always the force times the perpendicular distance to the line of action of the force.

Supplementary Exercises

Determine the components of the reactions for the beams in Exercises 2-1 through 2-10 shown below.

2-1.

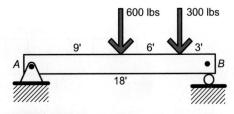

FIGURE E2-1

2-2.

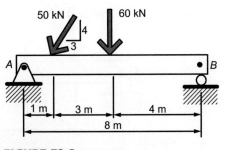

FIGURE E2-2

2-3.

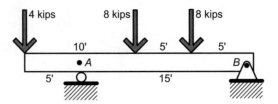

FIGURE E2-3

2-4.

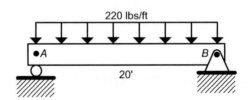

FIGURE E2-4

2-5.

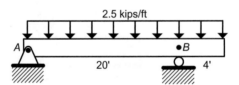

FIGURE E2-5

2-6.

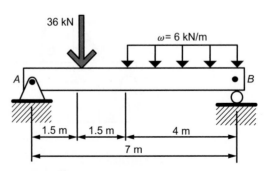

FIGURE E2-6

2-7.

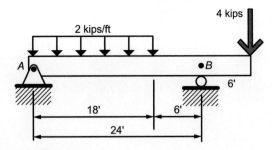

FIGURE E2-7

2-8.

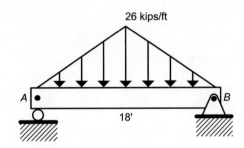

FIGURE E2-8

2-9.

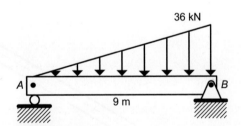

FIGURE E2-9

2-10.

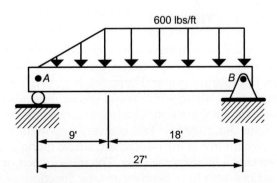

FIGURE E2-10

CHAPTER 3

Trusses

3.1 LINKS

A truss is a structural system composed of members called *links* or *two-force members*. A link is a rigid member attached to other members or supports by pinned connections. It is called a two-force member or *axially loaded member* since it can only be loaded in the longitudinal direction at the pinned connections as shown in Figure 3-1.

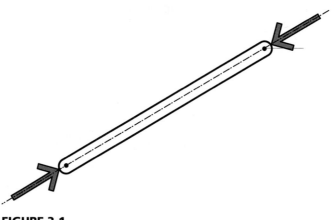

FIGURE 3-1

A link will have internal forces termed *link forces* that resist external forces. To verify the existence of these internal forces, or link forces, we can cut out a small piece of the link near the center (Figure 3-2, page 69). If the link was in equilibrium, then the small piece, along with the end pieces, must also be in static equilibrium. The link forces and the reactions in a two-force member follow the geometry of the member (Figure 3-3, page 69). This is a very important issue when analyzing trusses. If an external force is introduced anywhere along the length of the link, the member becomes *a three-force member*. The introduction of a force between the pinned connections introduces bending and the directions of the reactions are no longer known (Figure 3-4, page 69).

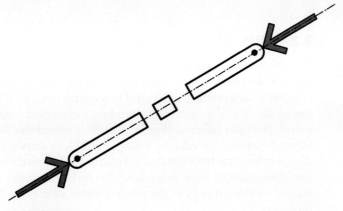

FIGURE 3-2

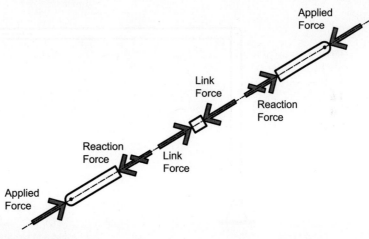

Applied
Force

Link
Force

Reaction
Force

Reaction
Force

Link
Force

Applied
Force

FIGURE 3-3

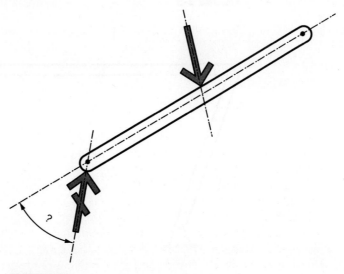

FIGURE 3-4

3.2 TRUSSES

If a structure is unstable, it will collapse when subjected to either gravity or lateral loads. If we create a structural system composed of links (Figure 3-5), what would happen if the system is subjected to a lateral load, *P*? Collapse? Since there is nothing to stabilize the system (remember, the pinned connection does not resist moment), it would collapse (Figure 3-6). To prevent this from happening, we can stabilize the system by adding a diagonal link as shown in Figure 3-7. By adding the diagonal link, we have created a truss. Now we can define a *truss* as a structural system composed of links in a triangular form. Since the truss is made up of links, it should be loaded only at the pinned joints, otherwise there will be bending in the members.

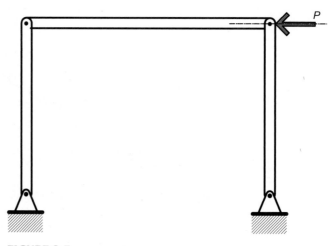

FIGURE 3-5

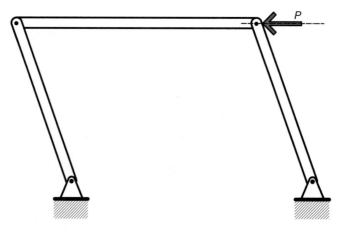

FIGURE 3-6

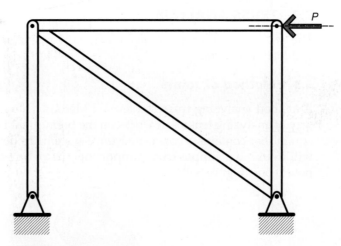

FIGURE 3-7

3.2.1 Terminology

The links or truss members that make up the perimeter of a truss are called *chords* (Figure 3-8). The members contained within the chords, the *verticals* and *diagonals*, are termed *web members*. The basic configurations of links that make up the span of the truss are called *panels*. We may also think of a panel as one or more smaller trusses that make up the larger truss.

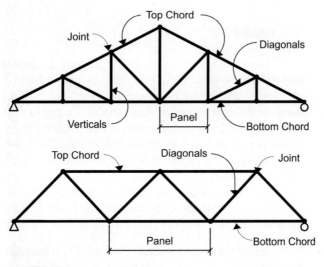

FIGURE 3-8

The joints are the points where the members meet. They may also be called *panel points* since they occur at the junction of the panels.

3.3 TRUSS ANALYSIS

The two common computational methods of analyzing trusses are *Method of Joints* and *Method of Sections*.

3.3.1 Method of Joints

A method analyzing trusses is termed Method of Joints. This is a relatively easy way of analyzing trusses. It works on the premise that each joint is a free-body diagram that contains a concurrent force system. To demonstrate this method, we will examine a simple truss supporting one of our gorillas, in this case a 300-pound one (Figure 3-9).

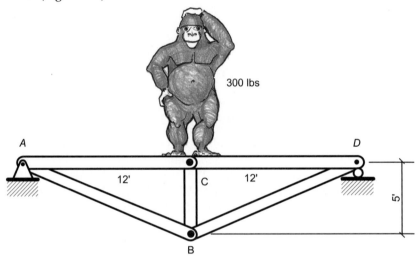

FIGURE 3-9

    As in any structural analysis problem, the first step is to examine the problem and determine what data are given. In this problem, we have a roller connection at support D and a pinned connection that has two unknowns, the x and y components of the reaction at A. A review of the applied load indicates that there are no horizontal forces; therefore, we can immediately assume that the sum of force in the x direction is zero and since our friendly gorilla is standing at mid-span, the vertical reactions must be equal to half of the gorilla's weight, 150 pounds. Note that the truss is composed of links, and link forces follow the geometry of the link indicating that the components of the link force will have the same slope as the truss member. As shown in Figure 3-10, the slope of the bottom chord is 5 in 12, which we will need to determine the components of the link force in these members.

    Before we begin to isolate the individual joints of the truss, we have to establish the following rule about link forces: *if the link force is pushing on the joint (arrow head pointing to the joint), the link is in compression; if the link force is pulling on the joint (arrow head pointing away form the joint), the link is in tension.* Once we determine that a member is in compression, then it must remain in compression throughout its length. A member that is in compression on one end cannot be in tension on the other.

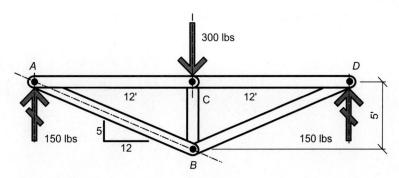

FIGURE 3-10

We can begin by examining Joint A. We can begin at any joint as long as we do not have more than three unknowns. We have three equations for static equilibrium and therefore we can only deal with three unknowns. If we were to select a joint with four or more unknowns, we would have to move to another joint where we have three or less unknowns and resolve some of the link forces, and then return to the previous joint.

At Joint A we have a horizontal link force in the top chord and the x and y component of the link force in the bottom chord AB. We have two horizontal unknowns (AC and AB_x) and only one vertical component (AB_y), so we will begin by summing forces in the y direction.

$$\Sigma F_y = 0$$
$$0 = 150 \text{ lbs} - AB_y$$
$$AB_y = 150 \text{ lbs}$$

Since the vertical component of AB must be acting down, AB_y must be negative (Figure 3-11).

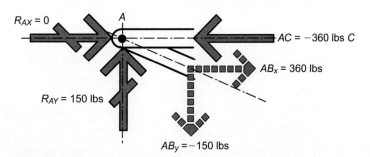

FIGURE 3-11

Now that we have AB_Y, we can use the geometry of AB (5 in 12 slope) to find AB_X and AB.

$$c = \sqrt{a^2 + b^2} = \sqrt{5^2 + 12^2} = 13$$

$$\frac{AB_Y}{5} = \frac{AB_X}{12} = \frac{AB}{13}$$

$$\frac{AB_y}{5} = \frac{150 \text{ lbs}}{5} = 30 \text{ lbs}$$

$$\frac{AB_x}{12} = 30 \text{ lbs}$$

$$AB_x = 12\,(30 \text{ lbs}) = 360 \text{ lbs}$$

$$\frac{AB}{13} = 30 \text{ lbs}$$

$$AB = 13\,(30 \text{ lbs}) = 390 \text{ lbs} \quad T$$

The link force in member AB is pulling on the joint and therefore is a tensile link force indicated by T after the numerical value. If the link force was pulling on the joint, the member would be in compression and C would appear after the numerical value (Figure 3-12).

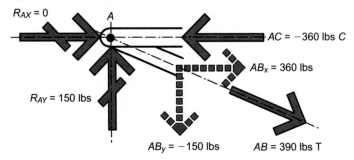

FIGURE 3-12

Since there are no horizontal forces acting on the truss, we can assume that $R_{AX} = 0$ and determine the link force in member AC.

$$\Sigma F_X = 0$$

$$0 = -360 \text{ lbs} + AC$$

$$AC = 360 \text{ lbs} \quad C$$

In this situation, the link force in AC is pushing on Joint A and therefore is a compressive link force.

Having completed Joint A, we can move to either Joint B or C. Joint C is the least complex, so we will proceed to investigate that joint (Figure 3-13). From Joint

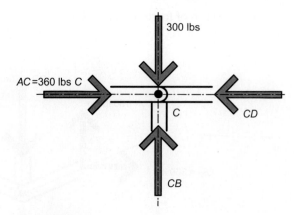

FIGURE 3-13

A we have the link force in member AC and we know the applied load of the gorilla's weight, 300 pounds. By summing forces in the x and y directions, we can readily determine the link forces in CD and CB.

$$\Sigma F_x = 0$$
$$0 = 360 \text{ lbs} - CD$$
$$CD = 360 \text{ lbs} \quad C$$

$$\Sigma F_y = 0$$
$$0 = 300 \text{ lbs} - CB$$
$$CB = 300 \text{ lbs} \quad C$$

Both CD and CB are pushing on Joint C and are compressive link forces (Figure 3-14).

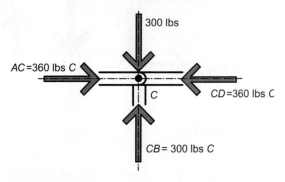

FIGURE 3-14

The only remaining joints are Joint B and Joint D. We can analyze either one and use the other to check our answers. If we have proceeded correctly, the last joint will have the sum of forces equal to zero without performing any mathematics. We have been working right to left, so we will choose Joint B (Figures 3-15 and 3-16).

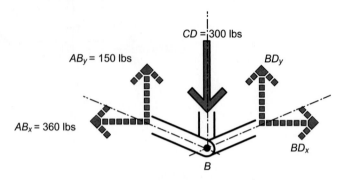

FIGURE 3-15

$$\Sigma F_y = 0$$

$$0 = -300 \text{ lbs} + 150 \text{ lbs} + BD_y$$

$$-BD_y = -150 \text{ lbs}$$

$$BD_y = 150 \text{ lbs}$$

$$\Sigma F_x = 0$$

$$0 = -360 \text{ lbs} + BD_x$$

$$-BD_x = -360 \text{ lbs}$$

$$BD_x = 360 \text{ lbs}$$

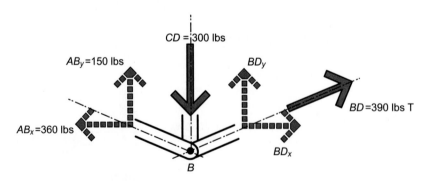

FIGURE 3-16

Having completed Joint B, we can place all of the link forces or components of the link forces on the last joint, Joint D, to check our work (Figure 3-17). Placing the values we calculated on the Joints B and C, we can immediately see that the joint is in static equilibrium and assume that our analysis is OK. We summarize our work as shown in Figure 3-18.

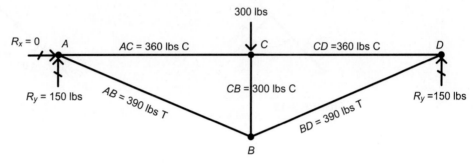

FIGURE 3-17

FIGURE 3-18

Even though they did not occur in this example, there are members that, under specific loading conditions, can have zero link force. It is important to recognize these members early in our investigation to simplify our analysis. These members usually occur in L- or T-shaped joints, which do not have applied forces or other link forces acting on them (Figure 3-19, page 78).

3.3.2 Method of Sections

The Method of Sections is based on the principle that if the entire structure is in static equilibrium, then all the parts of the structure must also be in static equilibrium. We cut a section through the truss knowing that if the truss is in equilibrium, the pieces we cut are in static equilibrium. The difficultly involved in this method is determining how to eliminate some of the unknown link forces, as we will see in the following example.

Our 300-pound gorilla is standing off center of the truss (Figure 3-20), so we will sum moments and forces to determine the reactions.

$$\Sigma M_A = 0$$

$$0 = -300 \text{ lbs } (6 \text{ ft}) + R_D (18 \text{ ft})$$

$$-R_D = \frac{-1,800 \text{ ft-lbs}}{18 \text{ ft}}$$

$$R_D = 100 \text{ lbs}$$

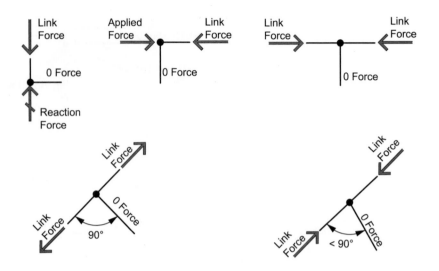

FIGURE 3-19

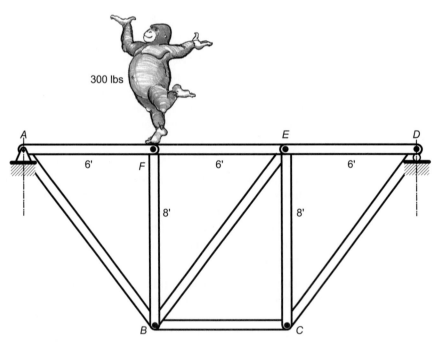

FIGURE 3-20

$$\Sigma F_y = 0$$

$$0 = -300 \text{ lb} + 100 \text{ lbs} + R_A$$

$$-R_A = -200 \text{ lbs}$$

$$R_A = 200 \text{ lbs}$$

Having the reactions and assuming we want the link forces in FE, EB, and BC, we can cut a section through the center panel of the truss (Figure 3-21).

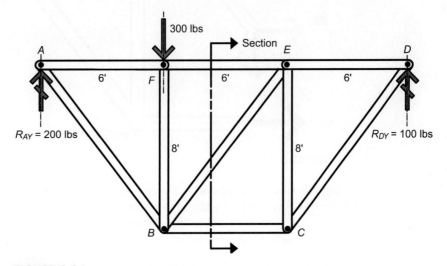

FIGURE 3-21

When using Method of Sections, we can only calculate the link forces in the members we cut with the section. The only applied forces we have to deal with are those that are acting on our section. We can ignore all the forces acting on the remaining portion of the truss. Either half of the truss could be selected, but selecting the right half allows us to eliminate the 300-pound force. The only forces we have to deal with are the force on the section we select, which is the reaction R_{DY} (Figure 3-22 , page 80).

We begin by reviewing the section and determining the unknown link forces, and components. Examining the vertical forces, we realize there is only one unknown component, EB_Y.

$$\Sigma F_Y = 0$$

$$0 = 100 \text{ lbs} - EB_Y$$

$$EB_Y = 100 \text{ lbs}$$

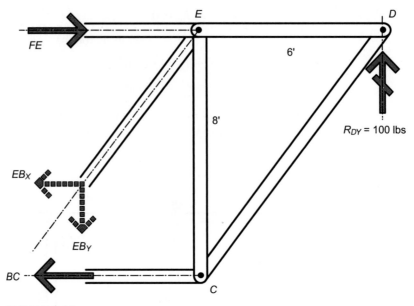

FIGURE 3-22

Having the vertical component, we can solve for the horizontal component, EB_x, and the link force in EB by using the geometry of the truss (the slope of EB is 6 in 8 or 3 in 4). We are now left with three horizontal unknowns and must look for a way to eliminate a few of these unknown values.

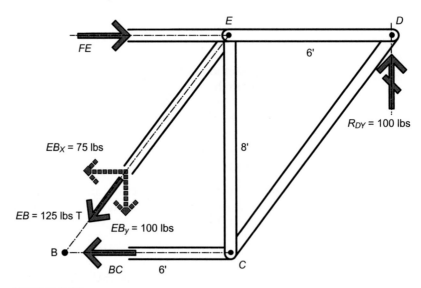

FIGURE 3-23

Figure 3-23 demonstrates that members EB and BC intersect at Joint B. Since the line of action of both these link forces extends through B, they cannot create moment about that point. Further, we can move any force along its line of action without changing its effect on the truss. We will move FE to Joint E.

$$\Sigma M_B = 0$$

$$0 = 100 \text{ lbs (12 ft)} - FE \text{ (8 ft)}$$

$$FE = \frac{1,200 \text{ ft-lbs}}{8 \text{ ft}} = 150 \text{ lbs} \quad C$$

Now with only one unknown link force, we can sum forces in the x direction to re-solve the balance of link forces in our section.

$$\Sigma F_x = 0$$

$$0 = FE - EB_x - BC$$

$$= 150 \text{ lbs} - 75 \text{ lbs} - BC$$

$$BC = 75 \text{ lbs} \quad T$$

Using the Method of Sections, we can cut the truss at any convenient locations. It does not have to go through a single panel but can, as illustrated in Figure 3-24, cut through two or more panels.

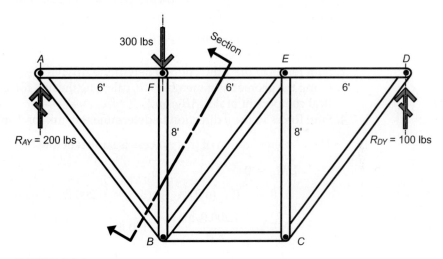

FIGURE 3-24

Using the left section, we can calculate the link forces in FE, FB, and AB. Since there are four unknowns, we have to look for a place to eliminate a few of these link forces from our calculations. From Figure 3-25 we determine that taking moments about Joint B will eliminate the link forces in members FB and AB, since the line of action of these link forces, as well as the 300-pound applied force, extend through Joint B. Once again, we can move the link force FE to Joint F (transmissibility).

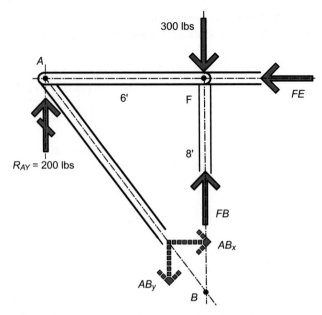

FIGURE 3-25

We will go through the following process in determining the link forces in this section:

1. Take moments about the Joint B. This will leave us with only one unknown horizontal component, AB_X;
2. Sum forces in the x direction to determine AB_X;
3. Using the geometry of member AB, calculate the link force in AB and the vertical component of AB, AB_X; and
4. Sum forces in the y direction to determine the link force in member FB.

The implementation of this process is shown below.

$$\Sigma M_B = 0$$

$$0 = -R_{A_Y}(6 \text{ ft}) + FE(8 \text{ ft}) = -200 \text{ lbs}(6 \text{ ft}) + FE(8 \text{ ft})$$

$$-FE = \frac{-1{,}200 \text{ ft-lbs}}{8 \text{ ft}} = -150 \text{ lbs}$$

$$FE = 150 \text{ lbs}$$

$$\Sigma F_x = 0$$

$$0 = -150 \text{ lbs} + AB_X$$

$$-AB_x = -150 \text{ lbs} = -150 \text{ lbs}$$

$$AB_x = 150 \text{ lbs}$$

$$c = \sqrt{a^2 + b^2} = \sqrt{6^2 + 8^2} = 10$$

$$\frac{AB}{10} = \frac{AB_Y}{8} = \frac{AB_x}{6}$$

$$\frac{AB_x}{6} = \frac{150}{6} = 25 \text{ lbs}$$

$$\frac{AB_Y}{8} = 25 \text{ lbs}$$

$$AB_y = 8 \, (25 \text{ lbs}) = 200 \text{ lbs}$$

$$\frac{AB}{10} = 25 \text{ lbs}$$

$$AB = 250 \text{ lbs} \quad T$$

$$\Sigma F_y = 0$$

$$0 = -AB_y - 300 \text{ lbs} + 200 \text{ lbs} + FB$$

$$= -200 \text{ lbs} - 300 \text{ lbs} + 200 \text{ lbs} + FB$$

$$-FB = -300 \text{ lbs}$$

$$FB = 300 \text{ lbs}$$

As mentioned previously, the positive answers indicate we have made the correct assumption concerning the direction of the link forces. If we made an error in our assumptions, the numerical answer would be correct but our assumed direction would be in error and we would have to reverse the direction of link force. If we assumed the member was in compression, pushing on the joint, we would, after obtaining a negative answer, realize that it is in tension, pulling on the joint. To sum forces, it would be an important reminder to also show the new direction on the section diagram (Figure 3-26).

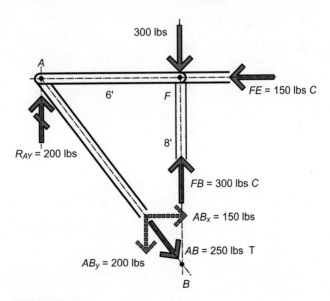

FIGURE 3-26

Sample Problems

Sample Problem 3-1: Determine the link forces in the truss shown in Figure S3-1 using the method of joints. The truss is subjected to wind loads that are perpendicular to the top chord.

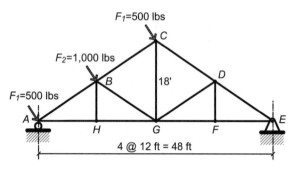

FIGURE S3-1

SOLUTION

Step 1 Rather than deal with forces acting at an angle with the horizon, we can begin this problem by resolving the applied forces into x and y components.

Since the truss chords are on a slope of 18 in 24 or 3 in 4, the applied forces therefore must be on a 4 in 3 slope (the slope of the force is rotated 90 degrees to the slope of the chord). We begin by determining the c side or hypotenuse.

$$c = \sqrt{a^2 + b^2} = \sqrt{4^2 + 3^2} = 5$$

$$\frac{F_1}{5} = \frac{F_{1y}}{4} = \frac{F_{1x}}{3}$$

$$F_{1y} = \frac{4F_1}{5} = \frac{4(500 \text{ lbs})}{5} = 400 \text{ lbs}$$

$$F_{1x} = \frac{3F_1}{5} = \frac{3(500 \text{ lbs})}{5} = 300 \text{ lbs}$$

$$\frac{F_2}{5} = \frac{F_{2y}}{4} = \frac{F_{2x}}{3}$$

$$F_{2y} = \frac{4F_2}{5} = \frac{4(1,000 \text{ lbs})}{5} = 800 \text{ lbs}$$

$$F_{1x} = \frac{3F_2}{5} = \frac{3(1,00 \text{ lbs})}{5} = 600 \text{ lbs}$$

Step 2 Now that we have the component forces, we can redraw the truss, substituting the component for the original forces. Once we have accomplished this, we can resolve the reactions at support A and E (Figure S3-2).

We will begin by solving for the x component of the reaction at E.

$$\Sigma F_x = 0$$

$$0 = -300 \text{ lbs} - 600 \text{ lbs} - 300 \text{ lbs} + R_{Ex}$$

$$R_{Ex} = 1{,}200 \text{ lbs}$$

$$\Sigma M_A = 0$$

$$0 = -600 \text{ lbs} (9 \text{ ft}) - 800 \text{ lbs} (12 \text{ ft}) - 300 \text{ lbs} (18 \text{ ft})$$

$$-400 \text{ lbs} (24 \text{ ft}) + R_{Ey} (48 \text{ ft})$$

$$R_{Ey} = 625 \text{ lbs}$$

$$\Sigma F_y = -400 \text{ lbs} - 800 \text{ lbs} - 400 \text{ lbs} + 625 \text{ lbs} + R_A$$

$$-R_A = -975 \text{ lbs}$$

$$R_A = 975 \text{ lbs}$$

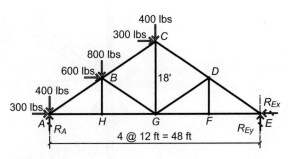

FIGURE S3-2

Step 3 Now that we have the components of forces F_1 and F_2 and the reactions, we can analyze the individual joints. We start our analysis by examining the truss to determine which members are zero-force members (Figure S3-3, page 86). We recognize the members BH, DF, and DG are zero-force members under this loading condition. Different loading conditions may make induce link forces into these members and create different zero-force members. In practical terms, these members are supporting member weights that are small and can be neglected.

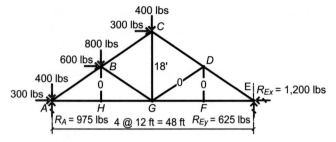

FIGURE S3-3

Step 4 We can start our analysis at any point, but it is convenient to start at a support such as Joint A, which has only one reaction, the vertical reaction (Figure S3-4). There is only one vertical unknown at Joint A, so we begin by summing the forces in the vertical or y direction.

$$\Sigma F_y = 0$$

$$0 = -400 \text{ lbs} + 975 - B_{A_y}$$

$$B_{A_y} = 575 \text{ lbs}$$

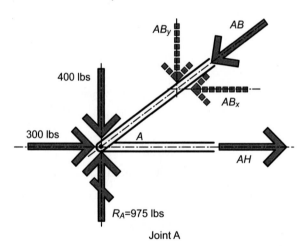

Joint A

FIGURE S3-4

B_{Ay} was assumed to be down at Joint A to oppose the reaction and therefore negative in the above equation. The answer was positive, which confirms that our assumption was correct. Realizing that the slope of AB is 18 in 24 or 3 in 4, we can proceed by resolving the link force in member BA and then move on to resolving member AH.

$$c = \sqrt{a^2 + b^2} = \sqrt{3^2 + 4^2} = 5$$

$$\frac{AB_x}{4} = \frac{AB}{5} = \frac{AB_y}{3} = \frac{575 \text{ lbs}}{3}$$

$$AB_x = \frac{4(575 \text{ lbs})}{3} = 766.67 \text{ lbs} \approx 767 \text{ lbs}$$

$$AB = \frac{5(575 \text{ lbs})}{3} = 958.33 \text{ lbs} \approx 958 \text{ lbs} C$$

The components of the link force are pushing on the joint; there-fore, the link force is pushing on the joint and must be in compression (Figure S3-5).

$$\Sigma F_x = 0$$

$$0 = 300 \text{ lbs} - 767 \text{ lbs} + AH$$

$$-AH = -467 \text{ lbs}$$

$$AH = 467 \text{ lbs} T$$

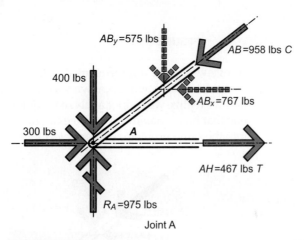

Joint A

FIGURE S3-5

Step 5 Since we already know that BH is a zero-force link, we can examine Joint H next. HG is the only remaining unknown and therefore must be equal and opposite of AH (Figure S3-6).

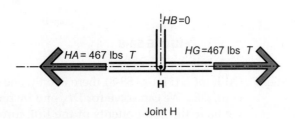

Joint H

FIGURE S3-6

Step 6 We will now proceed to Joint B as we work from left to right (Figure S3-7).

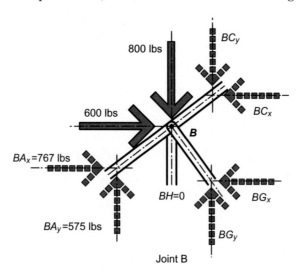

FIGURE S3-7

A review of Joint B informs us that there are two unknown horizontal or x components and two unknown vertical or y components. With this information we realize we cannot analyze this joint at this time and must move on to one that has a maximum of three unknowns. To ensure that we will not run into the same problem, we can select Joint E on the right end of the truss.

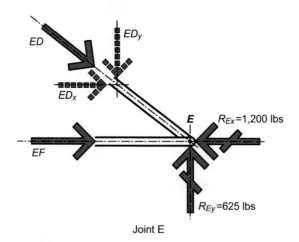

FIGURE S3-8

Step 7 On Joint E (Figure S3-8) there is only one unknown vertical or y component, DE_x. We can solve for DE_y and by proportion resolve the DE_x. Once we have the components of the link force using the sum of horizontal forces, we can calculate EF.

$$\Sigma F_y = 0$$

$$0 = -625 \text{ lbs} + DE_y$$

$$-DE_y = -625 \text{ lbs}$$

$$DE_y = 625 \text{ lbs}$$

$$\frac{DE_x}{4} = \frac{DE}{5} = \frac{DE_y}{3} = \frac{625 \text{ lbs}}{3}$$

$$DE_x = \frac{4(625 \text{ lbs})}{3} = 833.3 \text{ lbs} \approx 833 \text{ lbs}$$

$$DE = \frac{5(625 \text{ lbs})}{3} = 1,041.7 \text{ lbs} \approx 1,042 \text{ lbs} \quad C$$

$$\Sigma F_x = 0$$

$$0 = -1,200 \text{ lbs} + 833 \text{ lbs} + EF$$

$$-EF = -367 \text{ lbs}$$

$$EF = 367 \text{ lbs} \quad C$$

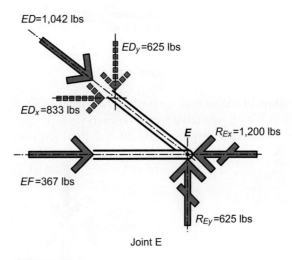

Joint E

FIGURE S3-9

Step 8 Having obtained the link forces for Joint E (Figure S3-9), we can now proceed to Joint D, transferring the link forces in DE to Joint D. Since members DF and DG are zero-force links, the link force in member DE must be equal to the link force in DE but opposite in direction (Figure S3-10).

Step 9 Having Joint D, we can proceed to analyze Joint F, transferring the link forces in EF in Joint E to Joint F. EF, being the only unknown force, must be equal and opposite to FG.

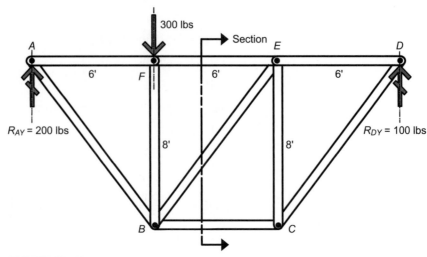

FIGURE S3-10

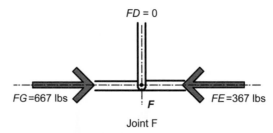

Joint F

FIGURE S3-11

Step 10 Now that we have Joint F (Figure S3-11) we can move to joint C. At Joint C we have two unknown vertical link forces but only one horizontal, so we can proceed with this joint (Figures S3-12).

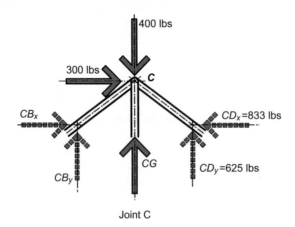

Joint C

FIGURE S3-12

$$\Sigma F_x = 0$$

$$0 = 300 \text{ lbs} - 833 \text{ lbs} + BC_x$$

$$-BC_x = -533 \text{ lbs}$$
$$BC_x = 533 \text{ lbs}$$

$$\frac{BC}{5} = \frac{BC_y}{3} = \frac{BC_x}{4} = \frac{533 \text{ lbs}}{4} = 133.25 \text{ lbs}$$
$$BC_y = 3(133.25 \text{ lbs}) = 399.75 \text{ lbs} \approx 400 \text{ lbs}$$
$$BC = 5(133.25 \text{ lbs}) = 666.25 \text{ lbs} \approx 666 \text{ lbs} C$$

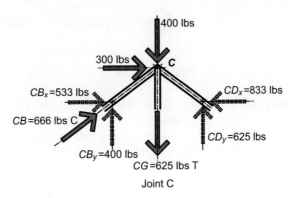

Joint C

FIGURE S3-13

Step 11 Joint G is the only remaining joint to be analyzed before returning to Joint B (Figure S3-14). At Joint G we have only one link force, BG, and we can resolve it by using components and summing forces. By proportions we can obtain BG (Figure S3-15).

$$\Sigma F_y = 0$$
$$0 = 625 \text{ lbs} - BG_y$$
$$BG_y = 625 \text{ lbs}$$

$$\frac{BG}{5} = \frac{BG_x}{4} = \frac{BG_y}{3} = \frac{625 \text{ lbs}}{3} = 208.33 \text{ lbs}$$

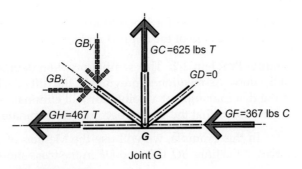

Joint G

FIGURE S3-14

$$BG_y = 3(208.33 \text{ lbs}) = 650 \text{ lbs}$$

$$BG = 5(208.33 \text{ lbs}) = 1,041.67 \text{ lbs} \approx 1,042 \text{ lbs}$$

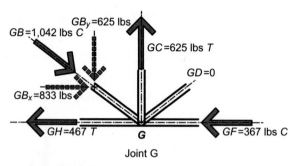

Joint G

FIGURE S3-15

Step 12 We can conclude this rather lengthy process by going back to Joint B and applying the balance of information we have obtained. If we have proceeded correctly, Joint B will be in static equilibrium. If not, we will have to find our error(s) (Figure S3-16).

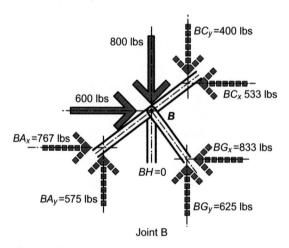

Joint B

FIGURE S3-16

Result

With the exception of a 1-pound error, due to rounding off our figures to whole numbers, the sum of the applied and link forces is zero.

Sample Problem 3-2: This sample problem demonstrates the use of the Method of Sections. As mentioned previously, the difficult issue in the use of this method is the determination of a point that will eliminate sufficient link forces when taking moments that the section becomes determinate.

In this example, we will use the Method of Sections to determine the link forces in the links BC, CD, and DE in the truss shown in Figure S3-17.

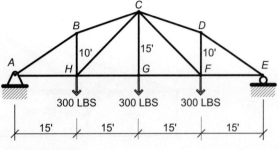

FIGURE S3-17

SOLUTION

Step 1 As usual, we begin by determining the reactions. The connection at Joint *A* is pinned and therefore can support both horizontal and vertical components. Since all the forces acting on the truss are vertical, the horizontal component of the reaction at *A* is zero. Realizing that the truss and the loads are symmetrical, we can also assume the reactions are equal and they are half of the total load or 450 pounds (Figure S3-18).

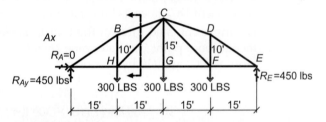

FIGURE S3-18

Having the reactions, we can cut our section and examine the forces acting on it. We only deal with those forces that act on the section and neglect all others (Figure S3-19). The section has four unknown forces or components, and therefore, we must find a point that will eliminate as many as possible.

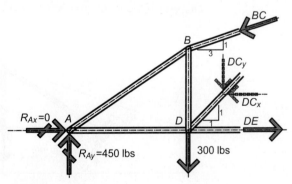

FIGURE S3-19

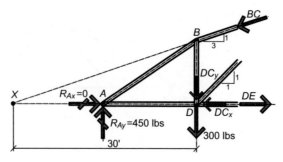

FIGURE S3-20

Step 2 It appears in Figure S3-20 that if we extend the top chord BC until it intersects the bottom chord, we will create a point x. This is a point where the lines of action of BC and DE meet and, as a result, cannot create moment. We can locate this point by using the slope of the top chord BC and the height of vertical, BD.

We further simplify the problem. Realizing that we can move a force or the components of a force anywhere along the line of action of the force, we move the components of DC to Joint D. This eliminates the horizontal component DC_x since its line of action now extends through the point x.

Step 3 Having eliminated three of the four unknowns, we proceed to take moments about point x to determine DC_y. Once this is accomplished we determine DC_x and DC by proportion.

$$\Sigma M_X = 0$$

$$0 = 450 \text{ lbs } (15 \text{ ft}) - 300 \text{ lbs } (30 \text{ ft}) - DC_y (30 \text{ ft})$$

$$DC_y = -75 \text{ lbs}$$

The negative answer tells us that our assumption that member DC is in compression is wrong. The negative sign indicates that the answer is numerically correct but the assumed direction is wrong. A graphic correction will be made before we proceed (Figure S3-21). Since DC has a slope of 1 in 1, we realize that DC_x equals DC_Y.

$$DC = \sqrt{DC_x^2 + DC_y^2} = \sqrt{(75 \text{ lbs})^2 + (75 \text{ lbs})^2} = 106 \text{ lbs } \quad T$$

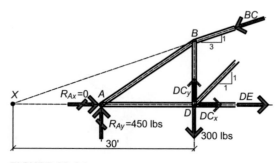

FIGURE S3-21

Step 4 Having determined *DC*, we will return to our original section and determine the balance of the unknown forces (Figure S3-22).

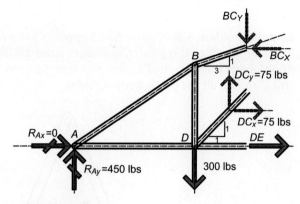

FIGURE S3-22

$$\Sigma F_y = 0$$

$$0 = 450 \text{ lbs} - 300 \text{ lbs} - 75 \text{ lbs} - BC_y$$

$$BC_y = 225 \text{ lbs}$$

$$\frac{BC_x}{3} = \frac{BC}{3.16} = \frac{BC_y}{1} = 225 \text{ lbs}$$

$$BC_x = 3(225 \text{ lbs}) = 675 \text{ lbs}$$

$$BC = 3.16(225 \text{ lbs}) = 711.51 \text{ lbs} \approx 712 \text{ lbs} \quad C$$

$$\Sigma Fx = 0$$

$$0 = 675 \text{ lbs} - 75 \text{ lbs} - DE$$

$$DE = 600 \text{ lbs} \quad T$$

We conclude this problem by placing these values on the section (Figure S3-23).

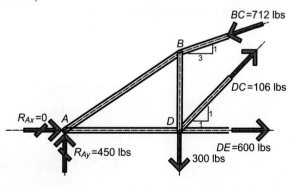

FIGURE S3-23

Note:

Taking moments about the point D provides an alternative method of analyzing this truss.

Sample Problem 3-3:   Figure S3-24 appears to be a truss but is in reality a structure composed of three-force members since all the members have forces introduced along their length. For this problem we have to determine the reactions at A and E and on all the pinned connections.

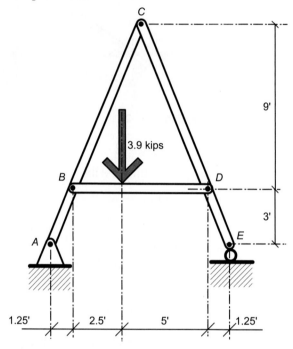

FIGURE S3-24

SOLUTION

Step 1 Determine the reactions at A and E by taking moments and summing forces. There are no horizontal forces acting on the system; therefore, we can assume $R_{Ax} = 0$.

$$\Sigma M_A = 0$$

$$0 = -3.9 \text{ kips } (3.75 \text{ ft}) + R_E (10 \text{ ft})$$

$$-R_E = -1.46 \text{ kips}$$

$$R_E = 1.46 \text{ kips}$$

$$\Sigma F_y = 0$$

$$0 = -3.9 \text{ kips} + 1.46 \text{ kips} + R_{Ay}$$

$$R_{Ay} = 2.44 \text{ kips}$$

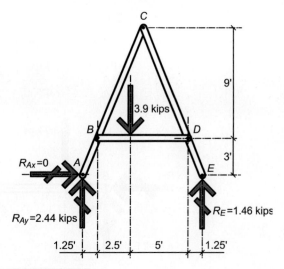

FIGURE S3-25

Step 2 Having the reactions (Figure S3-25), we can isolate the members of the system by using free-body diagrams. We start by isolating Member *BD* as a free-body diagram (Figure S3-26).

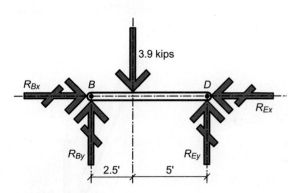

FIGURE S3-26

We have no knowledge of the relationship of the vertical components to the horizontal components, since *BD* is a three-force member. We can determine the vertical component of the reactions by taking moments.

$$\Sigma M_B = 0$$

$$0 = -3.9 \text{ kips } (2.5 \text{ ft}) + R_{Ey} (7.5 \text{ ft})$$

$$-R_{Ey} = \frac{-3.9 \text{ kips } (2.5 \text{ ft})}{7.5 \text{ ft}} = -1.3 \text{ kips}$$

$$R_{Ey} = 1.3 \text{ kips}$$

$$\Sigma F_y = 0$$

$$0 = -3.9 \text{ kips} + 1.3 \text{ kips} + R_{By}$$

$$-R_{By} = -2.6 \text{ kips}$$

$$R_{By} = 2.6 \text{ kips}$$

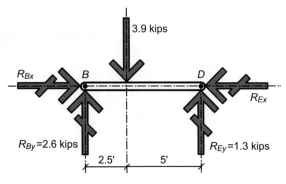

FIGURE S3-27

Step 3 Having the vertical components of BD (Figure S3-27), we can determine the horizontal components by using another free-body diagram, CE. Realizing the pin at D is in static equilibrium, we assume the vertical reaction at D on member BD is equal and opposite to the vertical reaction on member CE (Figure S3-28).

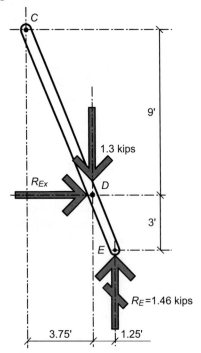

FIGURE S3-28

$$\Sigma M_C = 0$$

$$0 = 1.46 \text{ kips (5 ft)} - 1.3 \text{ kips (3.75 ft)} + R_{Ex} \text{ (9 ft)}$$

$$-R_{Ex} = 2.43 \text{ kips}$$

$$R_{Ex} = -2.43 \text{ kips}$$

Our answer is negative and therefore we have assumed the wrong direction for R_{Ex}. We will correct this assumption after we calculate the forces at the Joint C (Figure S3-29).

$$\Sigma F_y = 0$$

$$0 = 1.46 \text{ kips} - 1.3 \text{ kips} - R_{Cy}$$

$$R_{Cy} = .16 \text{ kips}$$

$$\Sigma F_x = 0$$

$$0 = 2.43 \text{ kips} - R_{Cx}$$

$$R_{Cx} = 2.43 \text{ kips}$$

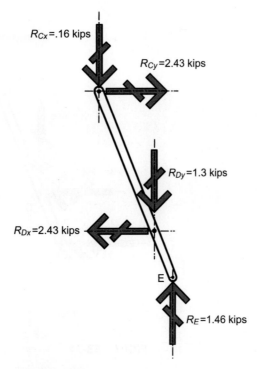

FIGURE S3-29

Step 4 Having the reaction, R_{Ex}, we can return to member BD to conclude our calculation on that member (Figure S3-30).

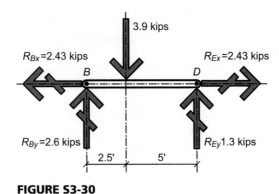

FIGURE S3-30

Step 5 We can use member AC to check our calculation by isolating it as a free-body diagram (Figure S3-31).

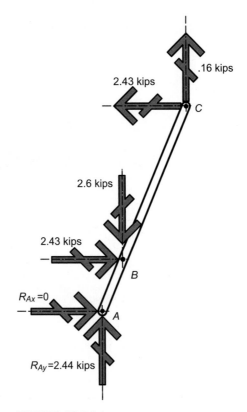

FIGURE S3-31

Results

Examination of member AC shows us that the sum of all the force in the x and y directions are equal to zero. We can conclude that our calculations are correct.

Supplementary Exercises

3-1. Using the Method of Joints, determine the link force in all of the members shown in the truss in Figure E3-1. Indicate whether the members are in compression or tension.

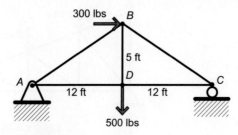

FIGURE E3-1

3-2. Determine the link forces in all the members of the truss shown in Figure E3-2 using the Method of Joints. Indicate whether the members are in compression or tension.

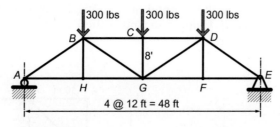

FIGURE E3-2

3-3. Using the Method of Joints, determine the link force in all of the members shown in the truss in Figure E3-3. Indicate whether the members are in compression or tension.

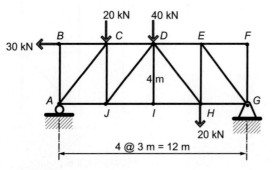

FIGURE E3-3

3-4. Using the Method of Joints, determine the link force in all of the members shown in the truss in Figure E3-4. Indicate whether the members are in compression or tension.

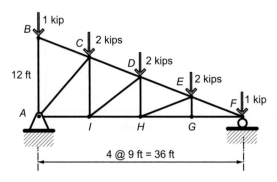

FIGURE E3-4

3-5. Verify your answers to Exercise 3-5 by determining the stress in CD, IH, and DI by the Method of Sections.

3-6. Using the Method of Sections, determine the stresses in members AB, AI, and IJ. Using a second section in Figure E3-5, calculate the stresses in CD, CG, and GH.

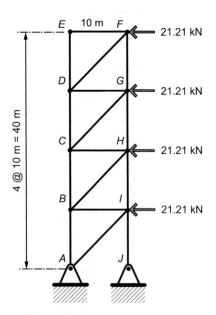

FIGURE E3-5

3-7. Figure E3-6 is a structure composed of two- and three-force members. Determine the reactions A and E and at all the pinned connections.

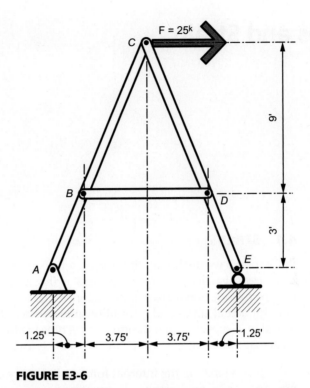

FIGURE E3-6

CHAPTER 4

Stresses and Strain

4.1 STRESS

In Chapter 3 we dealt with link forces, which are really the resultant of unit stress. *Unit stress*, usually referred to as *stress*, is the internal force, or link force, divided by the cross-sectional area of the member. The stresses are designated by the symbol f for actual stress and F for allowable stresses. These symbols are usually followed by a subscript that designates the type of stress. For example, f_t symbolizes an actual tension stress and F_t an allowable tension stress.

Again stress is the internal force or link force divided by the cross-sectional area and can be mathematically expressed as:

$$f \text{ or } F = \frac{P}{A}$$

where
  P is the internal force or link force in pounds, kips, or kilonewtons.
  A is the cross-sectional area in inches squared or meters squared.
  The units of stress are therefore expressed as lbs/in^2 (or psi), k/in^2 (or ksi), and kN/m^2 (Pa).

The symbol f is usually used for *structural analysis*, which is the process of determining the magnitudes of stresses in a given structure when subjected to known loads. For example, in Figure 4-1 we see that we have a known force of 350 pounds. If we assume that the vine has a diameter of 3/8 inch (.375 inch), we can calculate the cross-sectional area and tension stress in the vine using the formula above.

$$A = \frac{\pi d^2}{4} = \frac{\pi (.375 \text{ in})^2}{4} = .20 \text{ in}^2$$

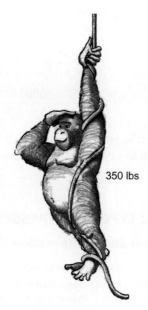

350 lbs

FIGURE 4-1

$$f_t = \frac{P}{A} = \frac{350 \text{ lbs}}{.20 \text{ in}^2} = \frac{1{,}783 \text{ lbs}}{\text{in}^2} = 1{,}783 \text{ psi}$$

The actual stress in the vine is 1,783 expressed in pounds per square inch (psi) because the units of the known force (the gorilla) are expressed in pounds.

F_t, the allowable stress, is usually used for *structural design*, the process of determining the size of a structural element so it can support given loads in a safe manner. Allowable stresses are found in structural design manuals such as the American Institute of Steel Construction (AISC) *Manual of Steel Construction*, the American Concrete Institute (ACI) *Building Code*, and American Forest and Paper Association (AF&PA) *National Design Specification for Wood Construction*.

Using Figure 4-1, we have a given load of 350 pounds. If we consult a structural design manual and found the allowable stress is 1,000 psi, we can size the diameter of vine required to safely support the gorilla.

$$A_{Req'd} = \frac{P}{F_t} = \frac{350 \text{ lbs}}{1{,}000 \text{ lbs/in}^2} = .35 \text{ in}^2$$

$$A = \frac{\pi d^2}{4}$$

$$d = \sqrt{\frac{4A}{\pi}} = \sqrt{\frac{4(.35 \text{ in}^2)}{\pi}} = .67 \text{ in}$$

$$\approx \frac{11}{16} \text{ in. dia.}$$

By examining the previous calculation we can determine the actual stress f_t, by dividing the load by the selected cross-sectional area.

$$A_{actual} = \frac{\pi d^2}{4} = \frac{\pi (11/16 \text{ in})^2}{4} = .37 \text{ in}^2$$

$$f_t = \frac{P}{A} = \frac{1{,}000 \text{ lbs}}{.37 \text{ in}^2} = 943 \text{ psi } < 1{,}000 \text{ psi } (F_t)$$

It is important to realize that to safely support an applied load the actual stress f must be equal to or less than the allowable stress, F.

$$f \leq F$$

4.2   BASIC TYPES OF STRESSES

4.2.1 Tension and Compression Stress

In the previous example we used actual and allowable tension stress. These stresses are normal stresses since they act perpendicular to the area of the cross section. Compression stresses act in a similar fashion as tension stresses, but in the opposite direction as shown in Figure 4-2. Tension and compression stresses are considered normal stress since they also act perpendicular, or normal, to the cross-sectional area.

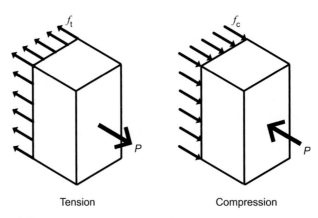

Tension   Compression

FIGURE 4-2

The analysis and design of members in compression is similar to those in tension as long as the compression members are short. Unlike tension stress, compression stresses are affected by the length of the member, an issue we will discuss further in Chapter 8. Tension stresses are unaffected by the length of the member.

To illustrate the calculation of compression stresses in a short post, we will use Figure 4-3. The known load is 350 pounds, the force of the gorilla's weight, and the cross section of the square post is 2 inches by 2 inches.

$$A = 2'' \times 2'' = 4 \text{ in}^2$$

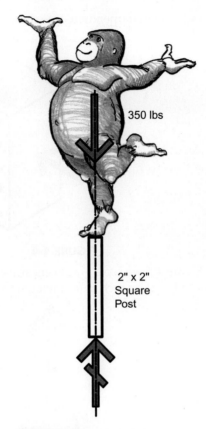

FIGURE 4-3

$$f_c = \frac{P}{A} = \frac{350 \text{ lbs}}{4 \text{ in}^2} = \frac{87.5 \text{ lbs}}{\text{in}^2}$$
$$= 87.5 \text{ psi}$$

To calculate the appropriate size of the post to support the force of the gorilla's weight, we choose a material that can support 50 psi or F_c, is 50 psi.

$$A_{req'd} = \frac{P}{F_c} = \frac{350 \text{ lbs}}{50 \text{ lbs/in}^2} = 7.0 \text{ in}^2$$
$$a = b = \sqrt{7.0 \text{ in}^2} = 2.65 \text{ in} \approx 2\tfrac{3}{4} \text{ in}$$

If we supply the gorilla with a post that is $2\tfrac{3}{4}$-inches square, he will be perfectly safe to perform acrobatics.

4.2.2 Shear Stress

A simple explanation of *shear stress* is the resistance of one plane of material to sliding over an adjacent plane. Shear stress is not a normal stress, because it acts parallel rather than perpendicular to the stressed area or cross section, as shown in Figure 4-4.

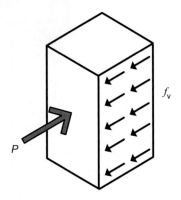

FIGURE 4-4

In Figure 4-5 we see a gorilla shearing a solid cylindrical piece of material. If the cylinder is $\frac{3}{4}$ of an inch in diameter, what is the actual shear stress?

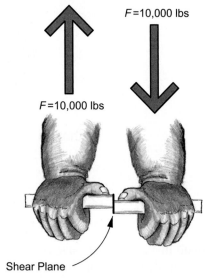

F=10,000 lbs

F=10,000 lbs

Shear Plane

FIGURE 4-5

Knowing the diameter, we can calculate the area and then the shear stress in the cylinder.

$$A = \frac{\pi d^2}{4} = \frac{\pi \left(2.5 \text{ in}^2\right)}{4} = 4.91 \text{ in}^2$$

$$f_v = \frac{P}{A} = \frac{10,000 \text{ lbs}}{4.91 \text{ in}^2}$$

$$= \frac{2,037 \text{ lbs}}{\text{in}^2} = 2,037 \text{ psi}$$

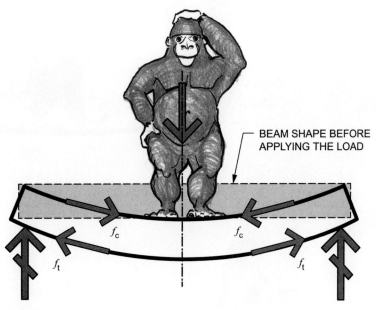

BEAM SHAPE BEFORE
APPLYING THE LOAD

f_c f_c

f_t f_t

FIGURE 4-6

4.2.3 Bending Stress

Bending stresses are a combination of tension and compression stresses. If we visualize a large gorilla standing in the center of a beam of somewhat less than adequate stiffness, we realize it would deflect (or move downward) as shown in Figure 4-6. We would also realize that as the beam bends and deflects, the ends of the member would rotate. The upper ends of the beam would rotate inward, indicating that the material on the top of the beam, is being compressed. On the bottom of the beam, the ends would rotate outward, indicating the material is being stretched or tensioned. While this will be explored in greater depth in Chapter 7, it is important for you to remember and understand the manner and type of stresses you can expect from a deflected beam.

4.3   STRESS/STRAIN

To simplify our explanations and calculations, we have assumed that our structures are rigid (except for the example in Figure 4-6). Clearly, structures are not rigid and will change shape when supporting loads. If a link in a truss is subjected to compression stresses, it is logical that it will shrink or shorten. The same holds true for a link that is subjected to tension stress; it will stretch or elongate. In most cases, the amount of shortening or lengthening is a function of the amount of stress and the type of material. The shortening or lengthening will be designated by the change in length, $\Delta \ell$. *Strain* is the ratio of $\Delta \ell$ to ℓ, the original length, and is denoted as ϵ. The units for strain are inches per inch, feet per foot, or meters per meter.

To illustrate the relationship of stress to strain, we will again call on our friendly gorillas. A gorilla weighing 350 pounds crosses a ravine on a thin, soft metal

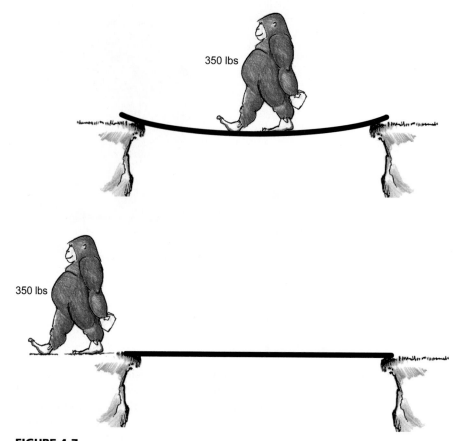

350 lbs

350 lbs

FIGURE 4-7

plate. While crossing, the plate deflects under the force of gorilla's weight. When the gorilla steps off the plate, it returns to its original shape (Figure 4-7 above).

Not long after, another gorilla weighing 500 pounds decides to follow the smaller gorilla and crosses the ravine on the same metal plate. This time, after the gorilla steps off the plate, the plate did not return to its original shape but remained in a deformed position (Figure 4-8 on next page).

How do we explain this phenomenon? In the 17th century, Robert Hook found the answer to this question. He discovered that most materials were essentially *elastic*, which means when the load was removed, they would return to their original shape. When the 500-pound gorilla walked across the plate, he exceeded the elastic limit of the material and permanently deformed the plate.

We can get a better understanding of material behavior under applied loads if we examine the results of a sample of soft steel that has been tension tested with the results graphed as Curve A of Figure 4-9. Here point *A* is the Proportional Limit. From zero up to point *A* the material exhibits a constant ratio of stress-to-strain (load-to-deformation). Until the stress on the test specimen reaches point *A*, it is in the *Elastic Range* and the material will return its original shape, when the load is removed. For the purposes of this text, all materials fall in this range. Once the

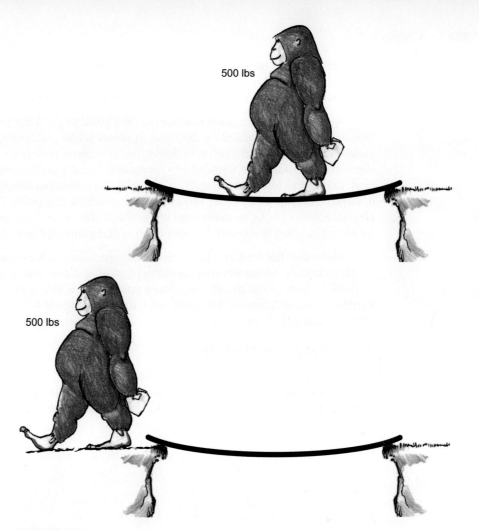

FIGURE 4-8

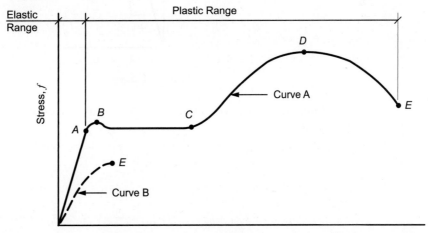

FIGURE 4-9

material has been stressed beyond this point, it enters the *Plastic Range,* which means it will never return to its original shape. In our example in Figure 4-8, the plate had reached its plastic range so it remained permanently deformed.

At point B the specimen reaches its *Yield Stress* (F_Y). After reaching the Yield Stress, the material exhibits a decrease in stress while continuing to elongate or yield. After this brief decrease in stress, the specimen continues to elongate at a somewhat constant rate without any increase in the stress. This yielding generates heat, causing strain hardening. As a result of the strain hardening at point C, the material accepts additional stress until it reaches the point D, which is the Ultimate Stress (F_u), the maximum stress the material can support. If the stress is increased beyond this point, the specimen will rupture at point D.

Materials that undergo large plastic deformation such as steel are considered *ductile* materials. Materials such as glass, plaster, and concrete do not exhibit plastic deformation. These materials rupture under load without much deflection and therefore are considered *brittle* material. Curve B in Figure 4-9 is an example of the stress/strain graph for a brittle material.

4.3.1 Modulus of Elasticity

In Figure 4-10 we have enlarged the beginning of Curve A in Figure 4-9 to clearly illustrate the linear stress-strain relationship of the material up to point A, the Proportional Limit. This ratio of stress to strain is called the *Modulus of Elasticity* (E).

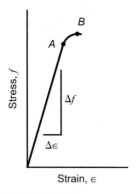

FIGURE 4-10

Mathematically we can express the modulus of elasticity in various forms as shown below.

$$E = \frac{f}{\epsilon}$$

$$= \frac{P/A}{\Delta \ell / \ell}$$

$$= \frac{P \ell}{\Delta \ell \, A}$$

The Modulus of Elasticity (E) is a measure of the materials stiffness. The strength of the material is a measure of the ordinates of Curve A, Figure 4-9. Since the units for strain are inches/inch, we can conclude that the Modulus of Elasticity will be the same as the units for stress—psi, ksi, and Pa. The approximate range of values for the Modulus of Elasticity for structural materials is from $.9 \times 10^6$ psi to 30×10^6 psi. Materials with high Modulus of Elasticity values are stiffer and offer greater resistance to deformation than materials with lower modulus of elasticity of the material.

Realizing that the Modulus of Elasticity is the ratio of stress to strain, we can manipulate the equation to obtain various values such as change of length or cross-sectional area. We will begin by using a material of known size and with the known Modulus of Elasticity ($E = 1.2 \times 10^6$ psi). In Figure 4-11, a force of 9,800 pounds is applied to the $3\frac{1}{2}$ inch $\times$ $3\frac{1}{2}$ inch test specimen. Using this data we want to determine the change in length ($\Delta\ell$) after the load is applied.

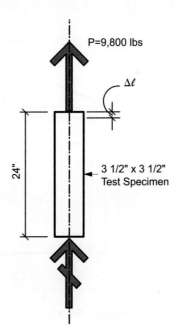

FIGURE 4-11

$$A = (31/2 \text{ in})^2 = 12.25 \text{ in}^2$$

$$f_c = \frac{P}{A} = \frac{9,800 \text{ lbs}}{12.25 \text{ in}^2} = \frac{800 \text{ lbs}}{\text{in}^2}$$

$$= 800 \text{ psi}$$

$$E = \frac{f}{\epsilon} = \frac{f}{\Delta\ell/\ell}$$

$$1.2 \times 10^6 \text{ lbs/in}^2 = \frac{800 \text{ lbs/in}^2}{\Delta\ell/24 \text{ in}}$$

$$\Delta\ell = \frac{24 \text{ in}(800 \text{ lbs/in}^2)}{1.2 \times 10^6 \text{ lbs/in}^2} = .016 \text{ in}$$

Sample Problems

Sample Problem 4-1: A large sign is supported by a $\frac{1}{4}$-inch diameter steel rod and a beam as shown in Figure S4-1. Assume a Modulus of Elasticity of 30×10^6 psi, and determine the stress in the rod and the elongation of the rod.

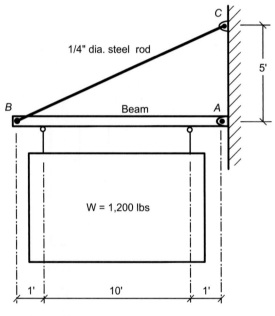

FIGURE S4-1

SOLUTION

Step 1 An analysis of the problem informs us that we must determine the reaction at B in order to determine the force in BC, the steel rod. The reaction will have to be supported by the vertical component of the link force in the rod. Since the sign supports are symmetrically located on the beam (Figure S4-2), we can safely assume the entire weight of the sign is located in the middle of the beam and they are half of the weight.

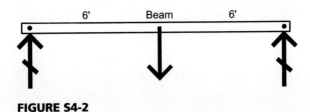

FIGURE S4-2

Step 2 The reaction at B must be supported by the rod; consequently, it becomes the vertical component of the link force. An examination of Figure S4-1 informs us that the rod must be in tension and therefore the vertical component must be up to pull on the joint at B. Now we can create a free-body diagram to determine the link force in the rod (Figure S4-3).

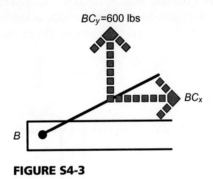

FIGURE S4-3

We can now find the horizontal component, F_x, and link force BC.

$$c = \sqrt{a^2 + b^2} = \sqrt{5^2 + 12^2} = 13$$

$$\frac{BC_X}{12} = \frac{BC}{13} = \frac{BC_Y}{5} = \frac{600 \text{ lbs}}{5} = 120 \text{ lbs}$$

$$BC_X = 12(120 \text{ lbs}) = 1{,}440 \text{ lbs}$$

$$BC = 13(120 \text{ lbs}) = 1{,}560 \text{ lbs}$$

Step 3 To determine the stress in BC, we can take the link force BC and divide it by the cross-sectional area of the rod.

$$A = \frac{\pi d^2}{4} = \frac{\pi(.25 \text{ in})^2}{4} = .05 \text{ in}^2$$

$$f_t = \frac{BC}{A} = \frac{1{,}560 \text{ lbs}}{.05 \text{ in}^2} = \frac{3{,}120 \text{ lbs}}{\text{in}^2} = 3{,}120 \text{ psi}$$

Step 4 To conclude this problem, we will determine the elongation of the rod using the Modulus of Elasticity.

$$E = \frac{f}{\epsilon} = \frac{f}{\Delta\ell/\ell}$$

$$\therefore \ \Delta\ell = \frac{f\ell}{E} = \frac{3,120 \text{ lbs/in}^2(13 \text{ ft})(12 \text{ in/ft})}{30 \times 10^6 \text{ lbs/in}^2} = .16 \text{ in}$$

Result

The stress in the rod is 3,120 psi and the total elongation of the rod under this load is .16 inches.

Sample Problem 4-2: A 10-foot high steel tube is subject to a load of 200 kips as shown in Figure S4-4. After the tube is loaded, the length changed by .00027 inches. What is the Modulus of Elasticity (E) of the steel?

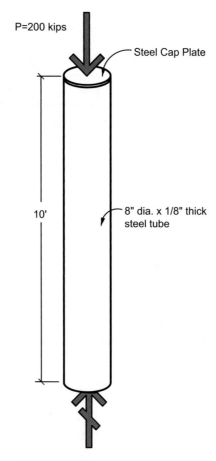

P=200 kips

Steel Cap Plate

10'

8" dia. x 1/8" thick steel tube

FIGURE S4-4

Step 1 Realizing the Modulus of Elasticity is the ratio of the stress to the strain, we begin this problem by calculating these values.

$$\epsilon = \frac{\Delta \ell}{\ell} = \frac{.27}{10'(12''/\text{ft})} = 2.25 \times 10^{-3}\,\text{in/in}$$

$$A = \frac{\pi d^2}{4} = \frac{\pi[(8\,\text{in})^2 - (7.75\,\text{in})^2]}{4} = 3.09\,\text{in}^2$$

$$f = \frac{P}{A} = \frac{200\,\text{kips}}{3.09\,\text{in}^2} = 64.72\,\text{ksi}$$

Using these values we can conclude our calculations by determining the Modulus of Elasticity.

$$E = \frac{f}{\epsilon} = \frac{64.72\,\text{ksi}}{2.25 \times 10^{-3}\,\text{in/in}} \approx 29 \times 10^3\,\text{ksi}$$

Result

The modulus of elasticity for this tube is 29×10^3 ksi, which is typical for most structural steel.

Supplementary Exercises

4-1. A $^3/_4$-inch diameter bolt is subjected to a tensile force of 12 kips as shown in Figure E4-1. What is the average shear stress in the bolt?

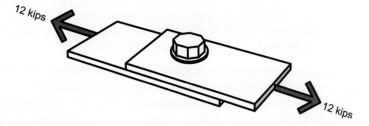

FIGURE E4-1

4-2. A concrete cylinder 18 inches high is subjected to a compression test. At 3,000 psi the strain is .003 inch/inch.
  a. What was the change in length of the cylinder at 3,000 psi?
  b. What is the modulus of elasticity of the concrete?
4-3. A steel column has an area of 15.6 in² and must support an axial load of 240 kips. If the column is 16 feet high, how much did the column shrink when the load was applied?
4-4. Determine the elongation of a 1-inch diameter steel rod, 10 feet long, used to support a 15 kip tensile force. Use $E = 29 \times 10^3$ ksi.
4-5. What load would be required to cause a steel column with a 3 in 2 cross section to shrink 0.25 inches if the column is 15 feet high? Use $E = 29 \times 10^3$ ksi.

CHAPTER 5

Load, Shear, and
Moment Diagrams

In order to select the appropriate material and cross section for a beam to support the external or applied forces, we must first examine the shear forces and moments acting on and within the beam. We will then be able to draw diagrams that graphically illustrate the shear forces and moments acting on the beam at any point along its length.

5.1 LOAD DIAGRAMS

The *load diagram* is nothing more than a graphic illustration of the initial shape of the beam (prior to applying any loads) with all external forces (applied forces and reactions) shown. To illustrate how to construct a load diagram, we will transform Figure 4-6 from Chapter 4 into a load diagram (Figure 5-1, next page).

To accomplish this, we will have to make a number of assumptions. We assume that our large gorilla weighs 400 pounds and is standing in the center of a 6-foot beam, which is simply supported. We will ignore the weight of the beam to keep the problem simple. Using these assumptions, the first phase of our load diagram would be illustrated as shown in Figure 5-2, next page. The only things missing to complete our load diagram are the reactions. Since the beam is symmetrically loaded, we can safely assume that the reactions are equal to half the load or 200 pounds each (Figure 5-3, next page).

5.2 SHEAR AND MOMENT

Having created our load diagram, we can now examine the shear stress and bending stress. We first cut a section through the beam at the point C, immediately to the right of the 400-pound load (Figure 5-4).

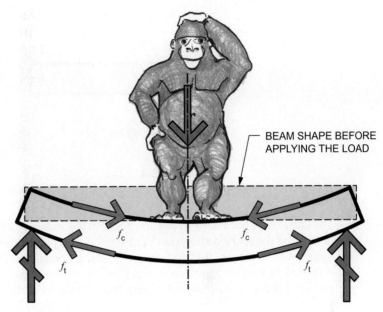

BEAM SHAPE BEFORE
APPLYING THE LOAD

f_c f_c

f_t f_t

FIGURE 5-1

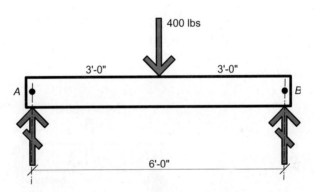

400 lbs

3'-0"   3'-0"

A   B

6'-0"

FIGURE 5-2

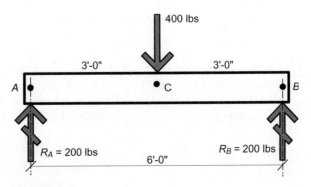

400 lbs

3'-0"   3'-0"

A   • C   B

R_A = 200 lbs R_B = 200 lbs

6'-0"

FIGURE 5-3

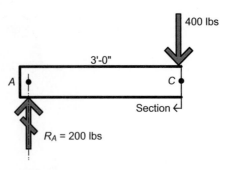

FIGURE 5-4

5.2.1 Shear

We know that the beam in Figure 5-3 is in static equilibrium and so we can conclude that the section shown in Figure 5-4 is also in static equilibrium. Upon closer examination we see that the sum of vertical forces is -200 pounds, not zero as required to keep the section in equilibrium. There must be an internal force that exists at the section that is keeping the section in equilibrium. This force is the resultant of a shear stress, which we will call V. We can determine the magnitude of this shear force by adding the external forces, F_E, acting on the section.

$$V = \Sigma F_E = 200 \text{ lbs} - 400 \text{ lbs} = -200 \text{ lbs}$$

If the sum of the external forces in the diagram is positive, the shear is positive. Since the sum of external forces in Figure 5-5 is negative, the shear must also be negative.

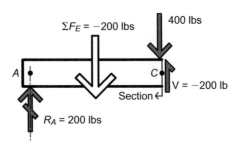

FIGURE 5-5

5.2.2 Moment

Now that we have the sum of forces on the section equal to zero, we can examine the section for moment. If we take moment about the point C in Figure 5-6, we realize the section would be rotating unless we have an internal resisting moment, M_R, to keep the beam in static equilibrium.

$$\Sigma M_C = -R_A \times 3 \text{ ft} + M_R$$
$$= -200 \text{ lbs} \times 3 \text{ ft} + M_R$$
$$M_R = 600 \text{ ft-lbs}$$

Now we have accounted for all the forces acting on the section as illustrated in Figure 5-6. The section is in static equilibrium.

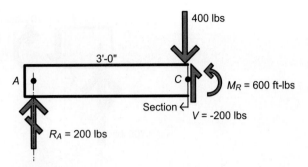

FIGURE 5-6

Unlike the moments created by external force that have their sign convention based on the direction of rotation, the internal resisting moment sign convention comes from shape of the deformed beam after loading. If the beam has a deformed shape of a bowl, the M_R is positive. If the deformed shape of the beam appears as an inverted bowl, then the M_R is negative (Figure 5-7).

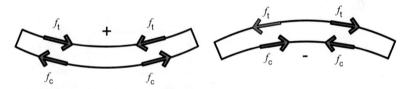

FIGURE 5-7

5.3 DIAGRAMING

To establish a system for graphing the shear and moment diagrams, we will take a series of sections through a few beams, starting with the beam shown in Figure 5-3. It is customary to take these sections from left to right. Our first section is taken immediately to the right of the reaction at A. Since the distance to the reaction is negligible, we can assume that there is no interior resisting moment on this section

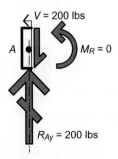

FIGURE 5-8

(Figure 5-8, on the previous page). Proceeding from left to right, we continue to taking sections at 1-foot intervals (Figure 5-9).

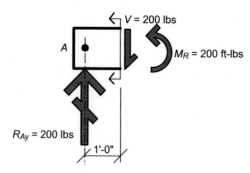

FIGURE 5-9

At 1 foot, 0 inch, there is no change in the external loads so the shear, V, remains 200 pounds but the moment, M_R, is now increased to 200 ft-lbs (Figure 5-10). At 2 feet, 0 inch, the shear remains 200 pounds, but once again the moment increases since the distance to the reaction increases. Our next section will be taken just shy of the applied load. This will allow us to see the effect of the concentrated load.

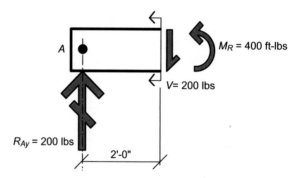

FIGURE 5-10

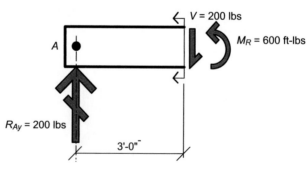

FIGURE 5-11

At the section shown in Figure 5-11 the shear remains unchanged and the moment increases to 600 foot-pounds. To complete our section at 3 feet, we will move our section immediately to the right of the applied load (Figure 5-12). We have only moved the section a very short distance from our previous section, but the magnitude of the internal resisting moment remains the same. By including the applied force of 400 pounds in the section, we changed the sum of the external forces, ΣF_E, from positive to negative, and thus the shear becomes negative. In the section shown in Figure 5-13, the magnitude of the moment has decreased and the shear remains the same. As we approach the right end of the beam, the moment continues to decrease but the shear remains unchanged (Figure 5-14).

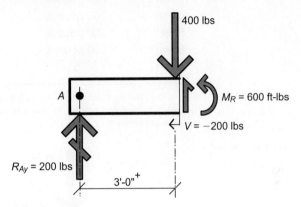

FIGURE 5-12

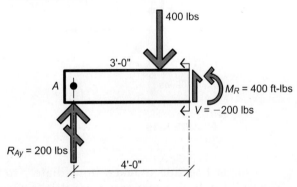

FIGURE 5-13

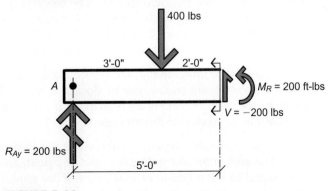

FIGURE 5-14

TABLE 5-1

Section	Shear	Moment
0	200 pounds	0
1 foot	200 pounds	200
2 feet	200 pounds	400
3 feet$^-$	200 pounds	600
3 feet$^+$	−200 pounds	600
4 feet	−200 pounds	400
5 feet	−200 pounds	200
6 feet*	−200 pounds	0
6 feet	0	0

*Note: Immediately to the left of the reaction (indicated by 6-feet in Table 5-1) the shear is −200 lbs. At 6 feet the reaction closes the diagram and brings the shear to zero.

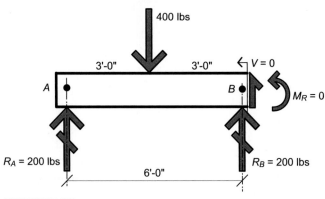

FIGURE 5-15

Our last section is taken immediately to the right of the reaction at B (Figure 5-15). In this section we see that the reaction brings the shear back to zero and brings the moment to zero. The beam remains in static equilibrium. To facilitate the graphing of the shear and moment diagrams, we will summarize the results of our sections in Table 5-1. Using the table values, we can now construct a shear and moment diagram (Figure 5-16).

It is hard to imagine that we would need to go through this lengthy process every time we want to analyze or design a beam. To shorten the process, let's examine the diagrams in Figure 5-16 to see if we can make a few observations that will help us to accelerate the process.

1. The maximum moment occurs where the shear is zero.
2. The area of the shear diagram and any point along the length of the beam is equal to the moment at that point. If we would like to find the moment at 2 feet from the left end of the beam, we can calculate it by determining the

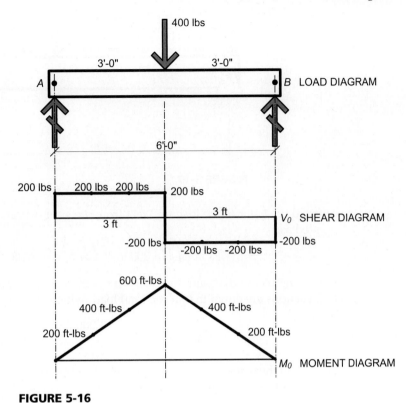

FIGURE 5-16

area of the shear diagram at 2 feet. Since this area is a rectangle, we will use the following formula:

$$A = xh$$

or

$$M = xV \text{ where } x \text{ is the distance form the left end of the beam}$$
$$= 200 \text{ lbs}(2 \text{ ft}) = 400 \text{ ft-lbs}$$

3. The ordinates (vertical heights) at any point along a diagram will equal the slope of the next higher diagram. For example, the ordinates of the shear diagram will equal the slope of the moment diagram. In Figure 5-16 the ordinates of the shear diagram (the vertical height from V_0 to the diagram) remain constant so the slope of the moment diagram is constant and must be a straight line.

Now that we have constructed a set of diagrams illustrating the shear and moments when a 400-pound gorilla stands in the center of a 6-foot long beam, we can follow the same procedure for the beam with a uniform distributed load of 200 pounds/foot. Once again, we will start by creating a load diagram from a graphic illustration (Figure 5-17 on next page). To do this, it will be necessary to calculate the reactions. W, the resultant of the uniform distributed load, will be located at the

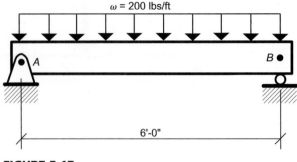

FIGURE 5-17

centroid or center of the load or 3 feet from the supports and is mathematically expressed as:

$$W = \omega L_L = 200 \text{ lbs/ft} \times 6 \text{ ft} = 1{,}200 \text{ lbs}$$

According to the diagram, the beam is symmetrically loaded, so we can assume the reactions are equal to half the total load, which is calculated as:

$$R_{Ay} = R_B = \frac{W}{2} = \frac{1{,}200 \text{ lbs}}{2} = 600 \text{ lbs}$$

Having all the external loads, we can complete our load diagram (Figure 5-18). Our first section is taken immediately to the right of the reaction at A (Figure 5-19 on next page).

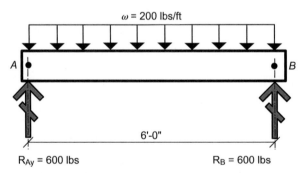

FIGURE 5-18

As we proceed with our section through the beam with uniform distributed load, it will be convenient for the purpose of calculating the shear and moment to determine W, the resultant of the uniform load on each specific section (Figures 5-20 through 5-24).

$$W = \omega L_L = 200 \text{ lbs/ft} \,(1 \text{ ft}) = 200 \text{ lbs}$$
$$V = \Sigma F_E = R_{Ay} - W$$
$$= 600 \text{ lbs} - 200 \text{ lbs} = 400 \text{ lbs}$$

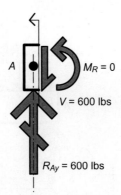

FIGURE 5-19

$$\Sigma M = 0 = R_A(1 \text{ ft}) - W(.5 \text{ ft}) + M_R$$
$$= -600 \text{ lbs}(1 \text{ ft}) + 200 \text{ lbs}(.5 \text{ ft}) + M_R$$
$$M_R = 500 \text{ ft} - \text{lbs}$$

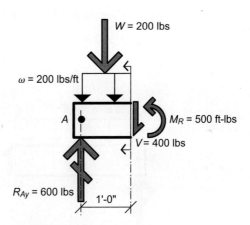

FIGURE 5-20

$$W = \omega L_L = 200 \text{ lbs/ft } (2 \text{ ft}) = 400 \text{ lbs}$$
$$V = \Sigma F_E = R_{Ay} - W$$
$$= 600 \text{ lbs} - 400 \text{ lbs} = 200 \text{ lbs}$$
$$\Sigma M = 0 = R_A(2 \text{ ft}) - W(1 \text{ ft}) + M_R$$
$$= -600 \text{ lbs}(2 \text{ ft}) + 400 \text{ lbs}(1 \text{ ft}) + M_R$$
$$M_R = 800 \text{ ft} - \text{lbs}$$

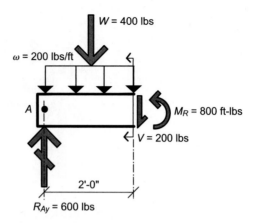

FIGURE 5-21

$$W = wL_L = 200 \text{ lbs/ft } (3 \text{ ft}) = 600 \text{ lbs}$$

$$V = \sum F_E = R_{Ay} - W$$

$$= 600 \text{ lbs} - 600 \text{ lbs} = 0$$

$$\sum M = 0 = R_A(3 \text{ ft}) - W(1.5 \text{ ft}) + M_R$$

$$= -600 \text{ lbs}(3 \text{ ft}) + 600 \text{ lbs}(1.5 \text{ ft}) + M_R$$

$$M_R = 900 \text{ ft} - \text{lbs}$$

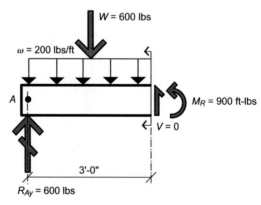

FIGURE 5-22

$$W = \omega L_L = 200 \text{ lbs/ft } (4 \text{ ft}) = 800 \text{ lbs}$$

$$V = \Sigma F_E = R_{Ay} - W$$

$$= 600 \text{ lbs} - 800 \text{ lbs} = -200 \text{ lbs}$$

$$\Sigma M = 0 = R_A(4 \text{ ft}) - W(2 \text{ ft}) + M_R$$

$$= -600 \text{ lbs}(4 \text{ ft}) + 800 \text{ lbs}(2 \text{ ft}) + M_R$$

$$M_R = 800 \text{ ft-lbs}$$

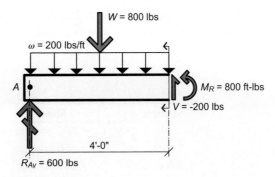

FIGURE 5-23

$$W = \omega L_L = 200 \text{ lbs/ft } (5 \text{ ft}) = 1,000 \text{ lbs}$$

$$V = \Sigma F_E = R_{Ay} - W$$

$$= 600 \text{ lbs} - 1,000 \text{ lbs} = -400 \text{ lbs}$$

$$\Sigma M = 0 = R_A(5 \text{ ft}) - W(2.5 \text{ ft}) + M_R$$

$$= -600 \text{ lbs}(5 \text{ ft}) + 1,000 \text{ lbs}(2.5 \text{ ft}) + M_R$$

$$M_R = 500 \text{ ft-lbs}$$

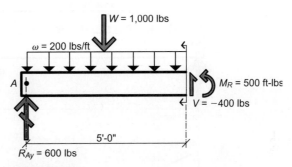

FIGURE 5-24

Our final section is taken immediately to the right of the reaction at B and includes R_B. The beam is in static equilibrium and the shear and moment are equal to zero (Figure 5-25).

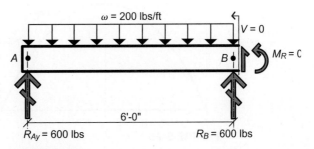

FIGURE 5-25

Having completed our sections, we will summarize the shear and moment values and graph a shear and moment diagram.

TABLE 5-2		
Section	Shear	Moment
0	600 pounds	0
1 foot	400 pounds	500 ft-lbs
2 feet	200 pounds	800 ft-lbs
3 feet	0	900 ft-lbs
4 feet	−200 pounds	800 ft-lbs
5 feet	−400 pounds	500 ft-lbs
6 feet*	−600 pounds	0
6 feet	0	0

*Note: Immediately to the left of the reaction (indicated by 6-feet in Table 5-2) the shear is −600 lbs. At 6 feet the reaction closes the diagram and brings the shear to zero.

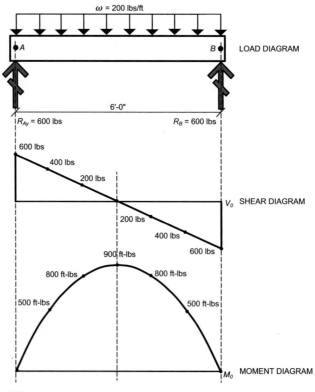

FIGURE 5-26

Having completed the diagrams for a uniform distributed load, we can test the observations that we made previously.

1. The maximum moment occurred at zero shear.
2. The area of the shear diagram and any point along the length of the beam is equal to the moment at that point. Let's check this by finding the moment at 2 feet from the left end of the beam. Since this area is a trapezoid, we will use the following formula:

$$A = \left(\frac{h_1 + h_2}{2} \right) b$$

$$M = \left(\frac{V_1 + V_2}{2} \right) b \quad \text{where } b \text{ is the distance from } V_1 \text{ to } V_2$$

$$= \left(\frac{600 \text{ lbs} + 200 \text{ lb}}{2} \right) 2 \text{ ft} = 800 \text{ ft-lbs}$$

3. The ordinates (vertical heights) at any point along a diagram will equal the slope of the next higher diagram. In Figure 5-26 the ordinates of the load diagram remain constant; thus, the shear diagram is a straight line. If we continue our inspection, we realize that the ordinates of the shear diagram are very high on the left and decrease in height until they reach zero at the center of the beam. These slopes result in the shape of the moment diagram being the geometric form called a *parabola*.

We can now further summarize the results of these diagrams in Table 5-3. Using our observations and Table 5-3, we can now prepare our load, shear, and moment diagrams without physically constructing sections. It is, however, helpful to imagine taking sections since we will continue to diagram the beams from left to right.

TABLE 5-3		
Beam load	**Shear diagram**	**Moment diagram**
Concentrated load	Horizontal lines	Sloped straight lines
Uniform distributed load	Sloped straight lines	Parabolic curved lines

These diagrams are not scaled drawings but rather accurate sketches. As mentioned in Chapter 1, our accuracy is increased by using grid paper when preparing the diagrams. To conclude this chapter, we will establish a process for graphing load, shear, and moment diagrams using Figure 5-27.

We begin by converting the illustration into a load diagram by calculating the reactions (Figure 5-28).

$$\Sigma M_A = 0$$

$$0 = -3 \text{ kips}(2 \text{ ft}) - 6 \text{ kips}(4 \text{ ft}) + R_B(6 \text{ ft})$$

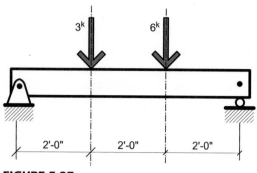

FIGURE 5-27

$$-R_B = \frac{-30 \text{ kip-ft}}{6 \text{ ft}} = 5 \text{ kips}$$

$$\Sigma F_y = -3 \text{ kips} - 6 \text{ kips} + 5 \text{ kips} + R_{Ay}$$

$$R_{Ay} = 4 \text{ kips}$$

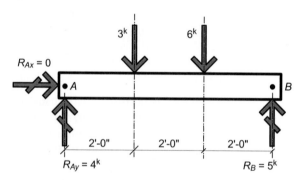

FIGURE 5-28

Having the load diagram, we can begin construction of the shear and moment diagrams by setting up vertical reference lines and horizontal zero shear and zero moment lines (Figure 5-29).

Now that we have our load diagram, we are ready to continue with the shear diagram. Again, we imagine taking sections from left to right. Our first section will be adjacent to the reaction at A. Note that the sum of external force is positive so the shear is positive or above our V_0 line. Proceeding from left to right with our imaginary sections, the shear remains constant until we come to the first concentrated load of -3 kips. At this point the shear is reduced from 4 kips to 1 kip. Continuing from left to right, we realize once again that there is no change in the shear until we reach the -6 kip concentrated load. At that point the shear is at 1 kip, and the force reduces, the shear to -5 kips. Since there is no additional external force, the shear remains -5 kips until it reaches the reaction at B, which closes the diagram at the end of the beam (Figure 5-30).

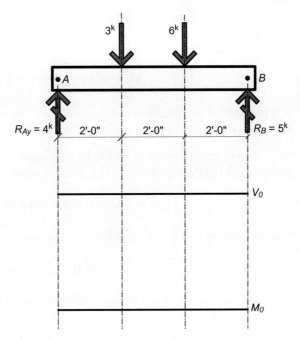

FIGURE 5-29

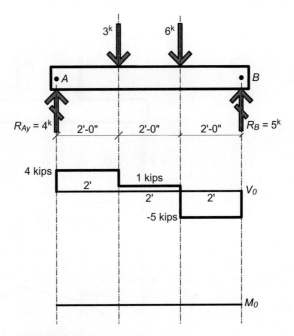

FIGURE 5-30

Since the beam is in static equilibrium, the moment diagram will begin at zero and we can calculate the moment at 2 feet by determining the area of the shear diagram from zero to 2 feet.

$$M_2 \text{ ft} = 4 \text{ kips}(2 \text{ ft}) = 8 \text{ kip-ft}$$

An inspection of the shear diagram shows the ordinates remain constant and therefore the moment diagram is a straight line (refer to Table 5-3).

The next moment is at 4 feet. This moment is the maximum moment, since it corresponds to the point where the shear is zero.

$$M_4 \text{ ft} = 8 \text{ kip-ft} + 2 \text{ kips} (2 \text{ ft}) = 10 \text{ kip-ft}$$

The last moment at 6 feet must bring the moment diagram back to zero.

$$M_{10} \text{ ft} = 10 \text{ kip-ft} + (-5 \text{ kips})(2 \text{ ft}) = 0$$

In many cases, if the moment diagram does not close or return to zero, it is due to mistakes in calculating the reactions. If the reactions are wrong but are equal to the sum of vertical forces, the shear diagram will close but the moment diagram will not (Figure 5-31). An examination of the moment diagram shows how the ordinates of the shear diagram affect the shape of the moment diagram. At the lowest slope of the moment diagram, from 2 feet to 4 feet, the ordinates of the

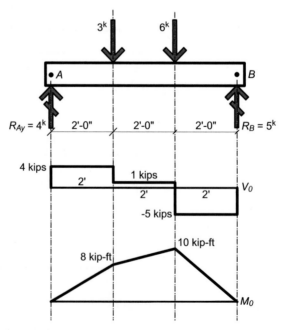

FIGURE 5-31

shear diagram are the smallest. The steepest slope exists from 4 feet to 6 feet where the ordinates of the shear diagram are the highest. The slope between 0 and 2 feet falls between the extremes since the ordinates fall between the lowest and the highest.

Sample Problems

Sample Problem 5-1: Draw the shear and moment diagrams for the simply supported beam shown in Figure S5-1.

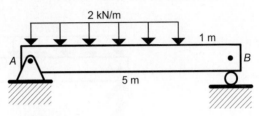

FIGURE S5-1

SOLUTION

Step 1 Determine the reactions to construct a load diagram (Figure S5-2).

$$W = \omega L_L = 2 \text{ kN/m}(4 \text{ m}) = 8 \text{ kN (at 2 m from } A)$$

$$\sum M_A = 0 = -8 \text{ kNm}(2 \text{ m}) + R_B(5 \text{ m})$$

$$R_B = \frac{16 \text{ kNm}}{5 \text{ m}} = 3.2 \text{ kN}$$

$$\sum F_y = 0 = 8 \text{ kN} - 3.2 \text{ kN} - R_{Ay}$$

$$R_{Ay} = 4.8 \text{ kN}$$

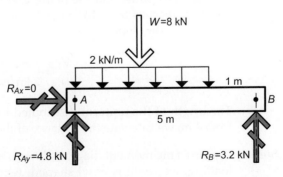

FIGURE S5-2

It is important to note that W, the resultant of the uniform distributed load, can be used to determine reactions but cannot be used for the shear and moment diagrams.

Step 2 Now that we have the load diagram, we can construct our guideline for the shear diagram (Figure S5-3). We begin by taking imaginary sections from left to right. Our first section is immediately to the right of the reaction at A. Since the sum of the external force is positive, the shear is positive and equal to R_{Ay} or 4.8 kN. Proceeding with our imaginary sections from left to right, the shear is decreasing at a constant rate of -2 kN/m for 4 meters where it reaches a value of -3.2 kN. From that point, on there is no change in the shear until R_B closes the diagram at the end of the beam.

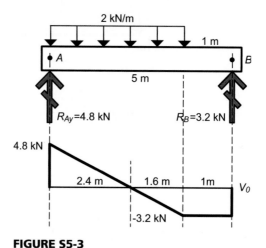

FIGURE S5-3

Referring to Table 5-3 for uniform distributed loads, the shear diagram has sloped straight lines. From 4 m to 5 m we realize that there is no change in the shear and the diagram remains a horizontal line until it reaches the reactions at B, which closes the diagram.

Realizing that we need the area of the shear diagram to calculate the moments, we must find the point, x, along the length of the beam where the shear reaches zero. To calculate this, we take the last shear value and divide it by the slope.

$$ x = \frac{V}{\omega} = \frac{4.8 \text{ kN}}{2 \text{ kn/m}} = 2.4 \text{ m} $$

The length of the uniform distributed load, L_L, is 4 m, subtracting 2.4 m from 4 m, we calculate the distance of the remaining triangle, 1.6 m.

Step 3 Construct the moment diagram. If we imagine the shape of the beam after loading, we see it is bowl shaped, which indicates the entire moment diagram is positive and above the M_0 line. Once again we refer to Table 5-3. The first portion of the moment diagram is parabolic in form and the last section is a sloped straight line. The straight line portion of the diagram must be tangent to the parabolic curve since they share a common ordinate on the shear diagram and thus a common slope.

To construct the moment diagram, we will need the maximum moment, the moment at 4 m, and the final moment to ensure the beam is in static equilibrium (Figure S5-4).

$$M_{max} = M_{2.4m} = \frac{1}{2}(4.8 \text{ kN})(2.4 \text{ m}) = 5.76 \text{ kN} - \text{m}$$

$$M_{4m} = 5.76 \text{ kN} - \text{m} + \frac{1}{2}(-3.2 \text{ kN})(1.6 \text{ m}) = 3.2 \text{ kN} - \text{m}$$

$$M_{5m} = 3.2 \text{ kN} - \text{m} - (3.2 \text{ kN})(1 \text{ m}) = 0$$

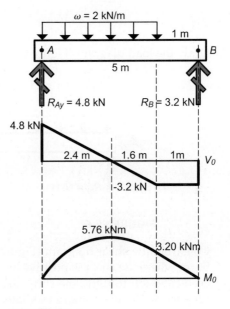

FIGURE S5-4

Results

The final moment returns to zero and thus ensures that we have the correct shear and moment.

Sample Problem 5-2: Draw the shear and moment diagrams for the beam with an overhang as shown in Figure S5-5.

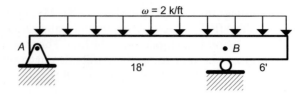

FIGURE S5-5

SOLUTION

Step 1 Determine the reactions.

$$W = \omega L_L = (2 \text{ kips/ft})(24 \text{ ft}) = 48 \text{ kips}$$

$$\sum M_A = 0 = -(48 \text{ kips})(12 \text{ ft}) + R_B(18 \text{ ft})$$

$$R_B = \frac{576 \text{ kip-ft}}{18 \text{ ft}} = 32 \text{ kips}$$

$$\sum F_y = 0 = -48 \text{ kips} + 32 \text{ kips} + R_{Ay}$$

$$R_{Ay} = 16 \text{ kips}$$

Step 2 Draw the load diagram (Figure S5-6).

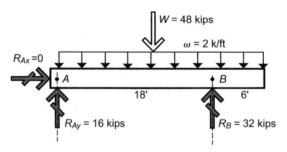

FIGURE S5-6

Step 3 Construct the shear diagram (Figure S5-7). We start at the left end of the beam where the shear is positive and equal to R_{Ay}, or 16 kips. As we proceed from left to right, the shear decreases at the rate of 2 kips/ft. By the time we arrive at a point immediately to the left of R_B, we have decreased 36 kips. Since we started at a positive 16 kips, the shear R_B is 16 kips – 32 kips or –20 kips. Once we move to the reaction R_B, the shear increases by 32 kips to 12 kips. The remaining load of –2 kips/ft on the 6-foot overhang brings the diagram back to zero.

To construct the moment diagram, we will need to know where the shear became zero.

$$x = \frac{V}{\omega} = \frac{16 \text{ kips}}{2 \text{ kips/ft}} = 8 \text{ ft}$$

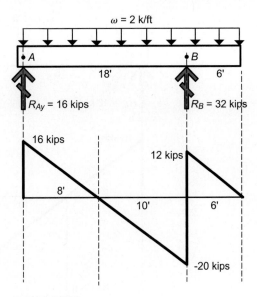

FIGURE S5-7

Step 4 Construct the moment diagram (Figure S5-8 on next page). Inspection of the shear diagram shows the shear reaches zero at two points, 8 feet and 18 feet. In this case we have two maximum moments, one positive and one negative. We begin our moment diagram by calculating the magnitude of the moments.

$$M_{8'} = \frac{1}{2}(16 \text{ kips})(8 \text{ ft}) = 64 \text{ kip-ft}$$

$$M_{18'} = 64 \text{ kip-ft} + \frac{1}{2}(-20 \text{ kip})(10 \text{ ft}) = -36 \text{ kip-ft}$$

$$M_{24'} = -36 \text{ ft} + \frac{1}{2}(20 \text{ kips})(6 \text{ ft}) = 0$$

Result

The final moment returns to zero to ensure us that we have the correct shear and moment.

Note:

Not having previously dealt with a beam with an overhang, it is interesting to review the moment diagram to see what we can learn. In cases where the load on the beam is a uniform distributed load, the crest or peak of the parabola is always up when the beam is subjected to gravity or downward loads (Figure S5-9, next page).

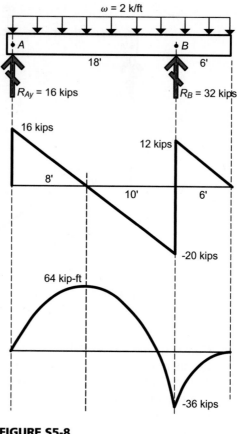

FIGURE S5-8

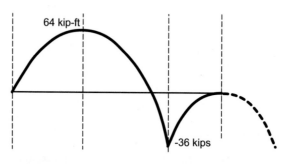

FIGURE S5-9

Finally, we have a point on all beams with overhangs, regardless of the loads, where the interior moment reaches zero. The point is termed a *point of inflection*. It demonstrates the relationship of the moment diagram to the deformed shape of the beam after loading (Figure S5-10). At the point of inflection, the moment changes sign. In this problem the moment changes from positive to negative and takes the exaggerated shape as shown in Figure S5-11.

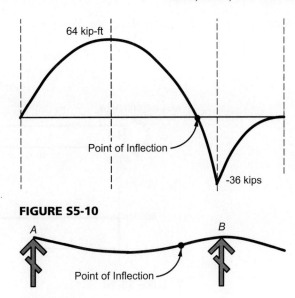

64 kip-ft

Point of Inflection

-36 kips

FIGURE S5-10

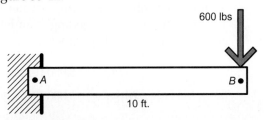

A

B

Point of Inflection

FIGURE S5-11

Sample Problem 5-3: Draw the shear and moment diagrams for the cantilevered beam shown in Figure S5-12.

600 lbs

A

B

10 ft.

FIGURE S5-12

SOLUTION

Step 1 To construct the load diagram, we must realize that the beam is cantilevered; therefore we have a moment reaction at A and only one vertical reaction since B is a free end. The reaction R_{Ay} must equal the load and therefore must be 600 pounds. To keep the beam in static equilibrium, the support at A must provide us with a moment equal to 600 pounds times 10 feet or 6,000 ft-lbs (Figure S5-13).

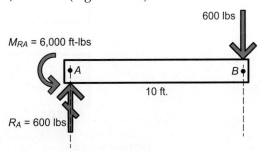

600 lbs

M_{RA} = 6,000 ft-lbs

A

B

10 ft.

R_A = 600 lbs

FIGURE S5-13

Step 2 Construct a shear diagram. We begin at the left end and proceed to the right. There is no change in the shear until we come to the concentrated load that closes the diagram (Figure S5-14).

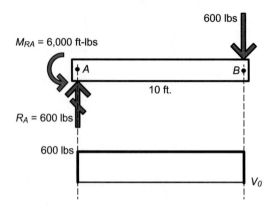

FIGURE S5-14

Step 3 Construct the moment diagram. We start with a moment reaction, M_{RA}, that is an applied moment rotating in a counterclockwise or positive direction. The internal resisting moment must be equal and opposite and therefore negative. Proceeding from left to right the moment increases at the rate of 600 pounds per foot until it reaches zero at the free end.

$$M_0 = -6,000 \text{ ft-lbs}$$

$$M_{10 \text{ ft}} = -6,000 \text{ ft-lbs} + 600 \text{ lbs}(10 \text{ ft}) = 0$$

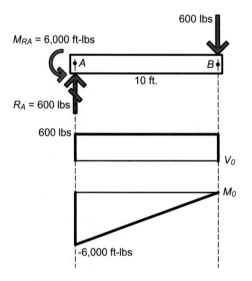

FIGURE S5-15

Results

The final moment returns to zero indicating the correct shear and moment have been calculated correctly.

Sample Problem 5-4: Draw the shear and moment diagram for the beam with mixed loads shown in Figure S5-16.

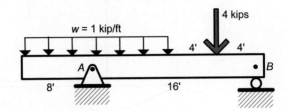

FIGURE S5-16

SOLUTION

Step 1   Calculate the reactions and draw the load diagram (Figure S5-17).

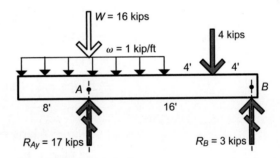

FIGURE S5-17

$$W = \omega L_L = (1\text{ kip/ft})(16\text{ ft}) = 16\text{ kips } (@8\text{ ft from } A)$$

$$\sum M_B = 0 = 16\text{ kips}(16\text{ ft}) + 4\text{ kips}(4\text{ ft}) - R_{AY}(16\text{ ft})$$

$$R_{AY} = \frac{272\text{ kip-ft}}{16\text{ ft}} = 17\text{ kips}$$

$$\sum F_Y = 0 = -16\text{ kips} - 4\text{ kips} + 17\text{ kip} + R_B$$

$$R_B = 3\text{ kips}$$

Step 2   Draw the shear diagram (Figure S5-18). At the free end (left end) of the beam, the shear is zero and as we move from left to right, the shear is decreasing at the rate of 1 kip/ft. We can calculate the shear at the important points in the following fashion:

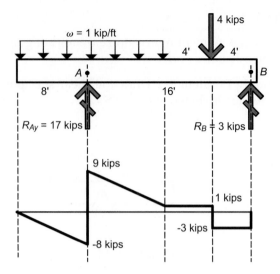

FIGURE S5-18

$$V_0 = 0$$

$$V_{8\,ft} = \omega L_L = -1\ \text{kip/ft}(8\ \text{ft}) = -8\ \text{kips}$$

$$V_{8\,ft} = -8\ \text{kips} + R_{Ay}$$

$$= -8\ \text{kips} + 17\ \text{kips} = 9\ \text{kips}$$

$$V_{16\,ft} = 9\ \text{kips} + \omega L_L$$

$$= 9\ \text{kips} + (-1\ \text{kip})(8\ \text{ft}) = 1\ \text{kip}$$

$$V_{20\,ft} = 1\ \text{kip}$$

$$V_{20\,ft} = 1\ \text{kip} - 4\ \text{kips} = -3\ \text{kips}$$

$$V_{24\,ft} = -3\ \text{kips} + R_B$$

$$= -3\ \text{kips} + 3\ \text{kips} = 0$$

Step 3 Construct the moment diagram (Figure S5-19). We will use the same method as previously employed to construct the moment diagram using the areas of the shear diagram.

$$M_0 = 0$$

$$M_{8\,ft} = \frac{1}{2}(-8\ \text{kips})(8\ \text{ft}) = -32\ \text{k-ft}$$

$$M_{16\,ft} = -32\ \text{k-ft} + \left(\frac{9\ \text{kips} + 1\ \text{kip}}{2}\right)8\ \text{ft} = 8\ \text{kip-ft}$$

$$M_{20\,ft} = 8\ \text{kip-ft} + 1\ \text{kip}(4\ \text{ft}) = 12\ \text{kip-ft}$$

$$M_{24\,ft} = 12\ \text{kip-ft} + (-3\ \text{kips})(4\ \text{ft}) = 0$$

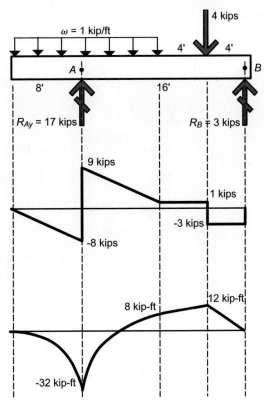

FIGURE S5-19

Results

The final moment returns to zero, indicating that the shear and moment have been calculated correctly.

Sample Problem 5-5: Draw the shear and moment diagram for the beam with a linear distributed load shown in Figure S5-20.

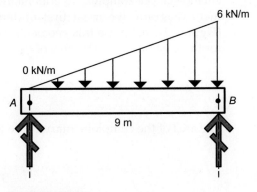

FIGURE S5-20

SOLUTION:

Step 1 Calculate W and the reactions. (See Figure S5-21.)

$$W = \frac{1}{2}\omega L_L = \frac{1}{2}(6 \text{ kN})(9 \text{ m}) = 27 \text{ kN } (@6 \text{ m from } A)$$

$$\sum M_A = 0 = -27 \text{ kN}(6 \text{ m}) + R_B(9 \text{ m})$$

$$R_B = \frac{162 \text{ kN} - \text{m}}{9 \text{ m}} = 18 \text{ kN}$$

$$\sum F_y = -27 \text{ kN} + 18 \text{ kN} + R_B$$

$$R_B = 9 \text{ kN}$$

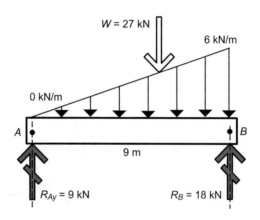

FIGURE S5-21

Step 2 Draw the shear diagram.

Inspection of the shear diagram shown in Figure S5-22 illustrates a shear diagram in the form of a parabolic curve, making mathematical calculations more complex. To understand the development of this type moment diagram, we must first understand the geometry of this form (Figure S5-23). To use this geometry we have to know the area equations for these geometries. The area of $\frac{1}{2}$ a parabolic curve is:

$$A = \frac{2}{3} bh$$

The area of the complementary curve is the remaining area or:

$$A = \frac{1}{3} bh$$

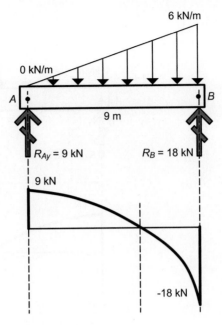

FIGURE S5-22

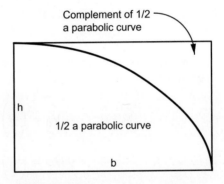

FIGURE S5-23

Using these equations, we can calculate the location of zero shear and the maximum moment (Figure S5-24, next page). Using similar triangles we can determine the w_{V_0}, the linear distributed load where the shear equals zero.

$$\frac{w_{V_0}}{xm} = \frac{6\ kN/m}{9\ m}$$

$$w_{V_0} = \left(\frac{6\ kN}{9\ m^2}\right) \times m$$

$$= \left(\frac{2\ kN}{3\ m}\right)x$$

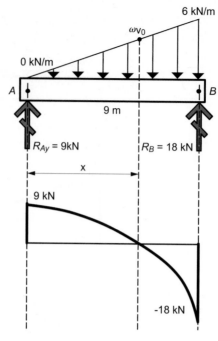

FIGURE S5-24

Now that we have an equation for ω_{V0}, we use the area of the load diagram to solve for x.

$$V_0 = \left(\frac{1}{2}\right)(\omega_{V_0})(x \text{ m})$$

$$9 \text{ kN} = \left(\frac{1}{2}\right)\left(\frac{2 \text{ kN}}{3 \text{ m}}x\right)(x \text{ m})$$

$$= \frac{1}{3}(x^2 \text{ kN})$$

$$9(3) = x^2$$

$$x = \mathbf{2}\ \overline{27} = 5.2 \text{ m}$$

Having the value of x, and recognizing the left portion of the shear diagram as $^1\!/_2$ a parabola, we can calculate the maximum moment using the area of the shear diagram.

$$M_{5.2 \text{ ft}} = \frac{2}{3}(9 \text{ kN})(5.2 \text{ m}) = 31.2 \text{ kN} - M$$

To determine if the moment diagram closes at 9 meters, we have to return to the geometry of the $^1/_2$ parabolic curve and incorporate our known structural values from above (Figure S5-25).

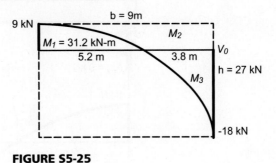

FIGURE S5-25

We must first calculate the moment for the rectangular area above V_0.

$$M = 9 \, \text{m}(9 \, \text{kN}) = 81 \, \text{kN-m}$$

To determine the moment below V_0, we need the remaining moment in the rectangular area M_2.

$$M_2 = 81 \, \text{kN-m} - 31.2 \, \text{kN-m} = 49.8 \, \text{kN} - M$$

To determine M_3, we have to calculate the moment of the complement $M_2 + M_3$.

$$M_1 + M_2 = \frac{1}{3}(27 \, \text{kN-m})(9 \, \text{m}) = 81 \, \text{kN-m}$$

We can conclude our calculations by determing M_3.

$$M_3 = (M_2 + M_3) - M_2 = 81 \, \text{kN-m} - 49.8 \, \text{kN} - \text{m} = 31.2\text{kN-m}$$

Results

The moment diagram closes on zero concluding that our calculations are correct (Figure S5-26).

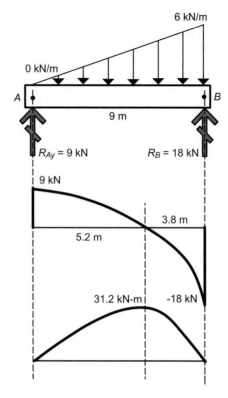

FIGURE S5-26

Supplementary Exercises

5-1. Construct the load, shear, and moment diagrams for Exercises 5-1 through 5-10.

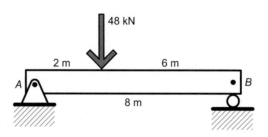

FIGURE E5-1

5-2.

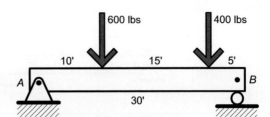

FIGURE E5-2

5-3.

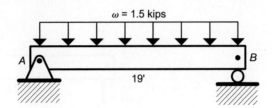

FIGURE E5-3

5-4.

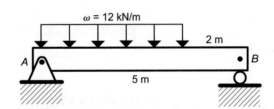

FIGURE E5-4

5-5.

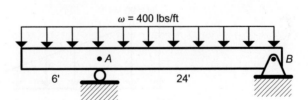

FIGURE E5-5

5-6.

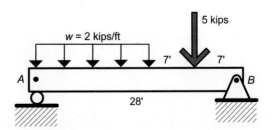

FIGURE E5-6

5-7.

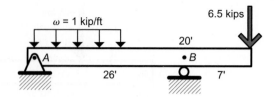

FIGURE E5-7

5-8.

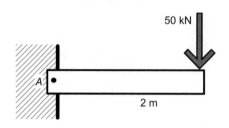

FIGURE E5-8

5-9.

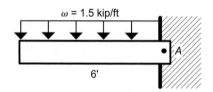

FIGURE E5-9

5-10.

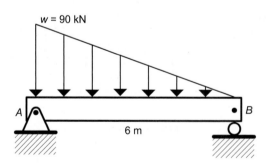

FIGURE E5-10

CHAPTER 6

Properties of a Cross Section

6.1 GENERAL

Now that we are able to calculate the shear and moment in beams, we have to determine how the cross section of a structural member contributes to the member's ability to support moment and shear. Two important properties of a cross section of a structural member are needed to determine the strength and stiffness of that member: the centroid and moment of inertia.

6.2 CENTROID

Most students are familiar with the term *center of gravity*. For example, suppose we have a square metal plate and that we wanted to suspend that plate from the ceiling by a single wire while keeping the plate level and parallel to the floor. To accomplish this, we would have to locate the center of gravity of the plate. If the plate had a uniform thickness, we would find the point of attachment (the center of gravity) at its *centroid* or center of area. We would then state that the *centroid* is the point where the area is concentrated.

The location of the centroid is described mathematically by placing the section in Cartesian coordinates, or on an x and y grid system. The distance from the x axis to the centroid is referred to as $\bar{y}$ and the distance from the y axis to the centroid is referred to as $\bar{x}$, as illustrated in Figure 6-1 on page 154.

If the section in Figure 6-1 has a width of b and a height of h, then we can logically conclude that $\bar{x} = b/2$ and $\bar{y} = h/2$. A sample of basic sections appears in Table 6-1 on page 154.

6.2.1 Centroid of Composite Sections

Composite sections are sections composed of basic shapes or pieces that have a centroid that we can readily calculate. A sample of these basic shapes is listed in

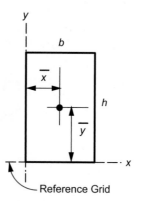

FIGURE 6-1

TABLE 6-1

Section	$\overline{x}$	$\overline{y}$	Area
Rectangle	$b/2$	$h/2$	bh
Triangle	$b/3$	$h/3$	$bh/2$
Circle	r	r	πr^2
Semicircle	r	$4r/3\pi$	$\pi r^2/2$
Quarter Circle	$4r/3\pi$	$4r/3\pi$	$\pi r^2/4$

Table 6-1. To determine the centroid of composite sections, we divide the sections into simpler shapes or pieces. We then, by use of Cartesian coordinates, establish a reference for x and y. To establish the centroid of the composite cross section, we use the following formula:

$$\bar{y} = \frac{\Sigma Ay}{\Sigma A}$$

or

$$\bar{x} = \frac{\Sigma Ax}{\Sigma A}$$

where y and x are the distances from the reference axis to the centroid of the pieces and A is the area of the piece.

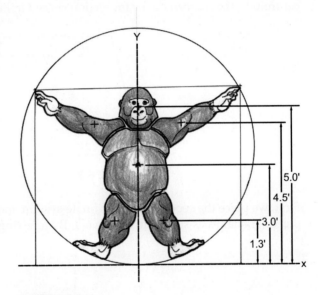

FIGURE 6-2

In Figure 6-2 we have the famous illustration of Leonardo Gorilla. This shows us a composite area that we will use to demonstrate the formulas. We will base our calculations on the dimensions given along with and following assumed areas:

Head = 1.2 square feet

Arms = 3.4 square feet

Body = 5.2 square feet

Legs = 3.0 square feet.

$$\Sigma A = 1.2 \text{ ft}^2 + 3.4 \text{ ft}^2 + 5.2 \text{ ft}^2 + 3.0 \text{ ft}^2 = 12.8 \text{ ft}^2$$

$$\Sigma Ay = 1.2 \text{ ft}^2 \, (5 \text{ ft}) + 3.4 \text{ ft}^2 \, (4.5 \text{ ft}) + 5.2 \text{ ft}^2 \, (3.0 \text{ ft})$$
$$+ \, 3.0 \text{ ft}^2 \, (1.3 \text{ ft}) = 40.8 \text{ ft}^3$$

$$\bar{y} = \frac{\Sigma Ay}{\Sigma A} = \frac{40.8 \text{ ft}^3}{12.8 \text{ ft}^2} = 3.2 \text{ ft}$$

It is important to understand that the location of the centroid for Leonardo is for this pose only. Should he put his arms down, the centroid would be lower than what we calculated. Because the gorilla is shown to be symmetrical about the y axis, the left half is a mirror image of the right half, it in unnecessary to calculate $\bar{x}$ since it would equal zero.

Having applied the centroid equations to a piece of art, let's now apply the centroid to the trapezoidal beam section shown in Figure 6-3.

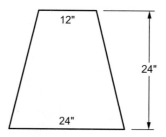

FIGURE 6-3

First, we place the section in a coordinate system and divide the section into basic pieces or shapes as shown in Figure 6-4. By referring to Table 6-1 for the area and

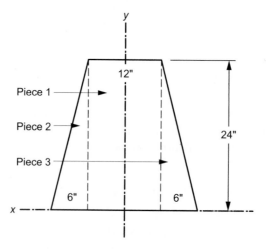

FIGURE 6-4

$\bar{y}$ for the pieces, we can set up a table to facilitate our calculations.

TABLE 6-2

Piece	Area (in²)	y (in²)	Ay (in³)
1	288	12	3,456
2	72	8	576
3	72	8	576
Σ	432		4,608

Using the last row of Table 6-2, we can calculate the centroid about the x axis.

$$\bar{y} = \frac{\Sigma Ay}{\Sigma A} = \frac{4,608 \text{ in}^3}{432 \text{ in}^2} = 10.67 \text{ in}$$

A review of Figure 6-4 shows the cross section is symmetrical about the y axis, so we can conclude that $\bar{x}$ is located on the y axis as shown in Figure 6-5.

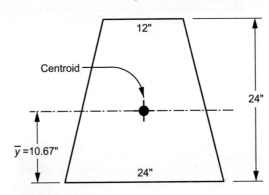

FIGURE 6-5

Next, we will calculate the centroid of the asymmetrical section shown in Figure 6-6.

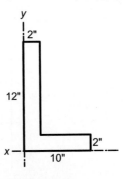

FIGURE 6-6

Once again, to calculate the centroid, we divide the section into pieces. Since the cross section is not symmetrical, we will have to calculate both $\bar{x}$ and $\bar{y}$. So we will expand our table to include $\bar{x}$. By referring to Table 6-1, we can complete Table 6-3.

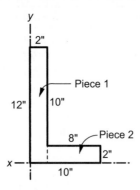

FIGURE 6-7

TABLE 6-3

Piece	Area (in²)	x (in)	Ax (in³)	y (in²)	Ay (in³)
1	24	1	24	6	144
2	16	6	96	1	16
Σ	40		120		160

Using row 3, the sums, we complete our calculations.

$$\bar{x} = \frac{\Sigma Ax}{\Sigma A} = \frac{120 \text{ in}^3}{40 \text{ in}^2} = 3 \text{ in}$$

$$\bar{y} = \frac{\Sigma Ay}{\Sigma A} = \frac{160 \text{ in}^3}{40 \text{ in}^2} = 4.0 \text{ in}$$

In this section the centroid exists in the space between the legs of the L shape (or angle). See Figure 6-8.

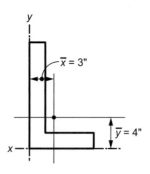

FIGURE 6-8

So why is knowing the location of centroid important in our study of structures? The answer can be found in Chapter 4. We determined that, when analyzing bending stresses, a simply supported beam was affected by compression stresses in the top portion and tension stresses in the bottom portion. These stresses changed from compression to tension at a horizontal plane located at the *centroid* and is called the *neutral axis*. The location of the centroid is essential to calculating the appropriate size of the structural member. This concept will be further developed next and in Chapter 7.

6.3   MOMENT OF INERTIA

Moment of inertia can be defined as the measurement of the strength and stiffness of a section based on the distribution of its area about the *neutral axis* or *centroid*.

The units of measurement of moment of inertia are inches to the fourth power (in^4). So, like our friendly gorilla (Figure 6-9), who is not calculus oriented, we can assume that this is an imaginary term. It is easy to imagine inches squared and inches cubed, but not so with inches to the fourth power.

FIGURE 6-9

We can help our understanding of this measurement by expanding Table 6-1 to include the moment of inertia or (I) of the basic or common shaped sections. (See Table 6-4 on page 160.) Using the equation for the moment of inertia for a rectangular cross section in Table 6-4, we can demonstrate how the moment of inertia is used to determine the strength and stiffness of structural members (Figure 6-10).

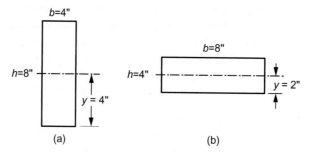

FIGURE 6-10

In Figure 6-10, section (*a*) is the same as section (*b*) except (*b*) has been rotated 90 degrees. Both sections (*a*) and (*b*) have the same area, but the areas are distributed differently. The area in section (*a*) is distributed in such a manner that its mass is further from the neutral axis than it is in (*b*). This results in a higher value for the moment of inertia.

Logic tells us that the cross section (*a*) is stiffer and stronger than cross section (*b*). Using the equation for *I* in Table 6-4, we can verify our assumption.

TABLE 6-4

Section	$\bar{x}$	$\bar{y}$	Area	I_x
Rectangle	$b/2$	$h/2$	bh	$bh^3/12$
Triangle	$b/3$	$h/3$	$bh/2$	$bh^3/36$
Circle	r	r	πr^2	$\pi r^4/4$ or $\pi D^4/64$
Semicircle	r	$4r/3\pi$	$\pi r^2/2$	$r^4(\pi/8-8/9\pi)$
Quarter Circle	$4r/3\pi$	$4r/3\pi$	$\pi r^2/4$	$r^4(\pi/16-4/9\pi)$

Section (a):
Locate the Neutral axis:

$$\bar{y} = \frac{h}{2} = \frac{8 \text{ in}}{2} = 4 \text{ in}$$

$$I_x = \frac{bh^3}{12} = \frac{4 \text{ in}(8 \text{ in}^3)}{12} = 107.67 \text{ in}^4$$

Section (b):
Locate the Neutral axis:

$$\bar{y} = \frac{h}{2} = \frac{4 \text{ in}}{2} = 2 \text{ in}$$

$$I_x = \frac{bh^3}{12} = \frac{8 \text{ in}(4 \text{ in}^3)}{12} = 42.67 \text{ in}^4$$

From these calculations we can conclude that section (a) has a larger I_x and therefore is stiffer and stronger than section (b).

6.3.1 Moment of Inertia of Composite Sections

In some cases it is convenient to calculate a composite section using a positive and a negative moment of inertia. An example of this type of section is a hollow rectangular section (Figure 6-11).

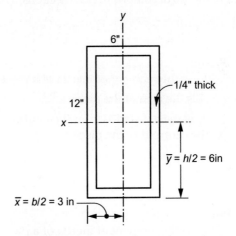

FIGURE 6-11

Since this section is symmetrical about both axes, we can assume that $\bar{y} = 6$ and $\bar{x} = 3$ inches. We then proceed to calculate the positive outer 6 inch $\times$ 12 inch rectangle minus the negative inner $5\frac{1}{2}$ inch $\times$ $11\frac{1}{2}$ inch rectangle.

$$I_x = I_{outside} - I_{inside}$$

$$I_x = \frac{bh^3}{12} - \frac{bh^3}{12}$$

$$= \frac{6 \text{ in}(12 \text{ in})^3}{12} - \frac{5.5 \text{ in}(11.5 \text{ in})^3}{12}$$

$$= 166.93 \text{ in}^4$$

$$I_y = I_{outside} - I_{inside}$$

$$I_y = \frac{bh^3}{12} - \frac{bh^3}{12}$$

$$= \frac{12 \text{ in}(6 \text{ in})^3}{12} - \frac{11.5 \text{ in}(5.5 \text{ in})^3}{12}$$

$$= 56.6 \text{ in}^4$$

Once again we are able to see how the distribution of area affects the moment of inertia. The top and bottom (in I_x calculation), even though they are smaller in area, are farther away from the neutral axis $\bar{y}$ than the sides are from the neutral axis, $\bar{x}$, yet I_x is almost three times larger than I_y.

The transfer formula provides us with another method of determining the moment of inertia of composite cross sections. The formula for the composite section is:

$$I_x = \Sigma I_{xc} + \Sigma A d_y^2$$

Where:

I_{cx} is the moment of inertia of a piece about its centroidal x axis,

A is the area of the piece, and

d_y is the y distance from the neutral axis ($\bar{x}$) of the composite section to the centroid of the piece.

or

$$I_y = \Sigma I_{cy} + \Sigma A d_x^2$$

Where:

I_{cy} is the moment of inertia of a piece about its y axis centroid,

A is the area of the piece, and

d_x is the x distance from the neutral axis ($\bar{y}$) of the composite section to the centroid of the piece.

To demonstrate this formula, we will compute the moment of inertia (I_x) of the composite cross section shown in Figure 6-12 on page 163. To determine I_x of

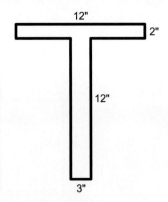

FIGURE 6-12

this cross section, we must first determine the location of the neutral axis or $\bar{y}$. As we have done previously, we will break the composite into pieces of known centroid and place the composite section on the x axis (Figure 6-13).

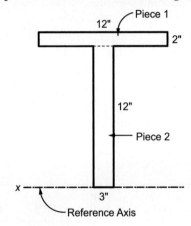

FIGURE 6-13

Using Table 6-5 we can locate $\bar{y}$.

TABLE 6-5			
Piece	**A (in²)**	**y (in)**	**Ay (in³)**
1	24	13	312
2	36	6	216
Σ	60		528

With the sum values in the third row of the table we can calculate $\bar{y}$, the distance to the neutral axis of the composite section (Figure 6-14).

$$\bar{y} = \frac{\Sigma Ay}{\Sigma A} = \frac{528 \text{ in}^3}{60 \text{ in}} = 8.8 \text{ in}$$

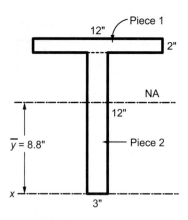

FIGURE 6-14

Now that we have located the neutral axis, we will calculate the moment of inertia of the component sections.

$$I_{x1} = \frac{bh^3}{12} = \frac{12 \text{ in}(2 \text{ in})^3}{12} = 8 \text{ in}^4$$

$$I_{x2} = \frac{bh^3}{12} = \frac{3 \text{ in}(12 \text{ in})^3}{12} = 432 \text{ in}^4$$

Having the I_x of the pieces, we can expand our table to include the moment of inertia (Table 6-6).

TABLE 6-6

Piece	Area (in^2)	y (in)	Ay (in^3)	I_{cx} (in^4)	d (in)	d^2 (in^2)	Ad^2 (in^4)	$I_c + Ad^2$ (in^4)
1	24	13	312	8	4.2	17.64	423.4	431.4
2	36	6	216	432	2.8	7.84	282.2	714.2
Σ	60		528					1,145.6

The $\Sigma(I_c + Ad^2)$ listed in the lower right-hand corner of the table provides us with the I_x for the composite area, 1,145.60 in^4.

When we calculate d for the table, it is important to realize that it is the distance from the neutral axis of the composite section to the centroid of the piece. With this in mind we can say d is the absolute value of $\bar{y}$ minus y.

$$d = |\bar{y} - y|$$

for example in Table 6-6:

$$d_1 = |8.8 \text{ in} - 13 \text{ in}| = 4.2 \text{ in}$$

$$d_2 = |8.8 \text{ in} - 6 \text{ in}| = 2.8 \text{ in}$$

With this specific section, we are also able to use the positive I_x plus the negative I_x that we used to find the moment of inertia of the hollow rectangular section in Figure 6-11. First, we will verify the location of the neutral axis, $\bar{y}$.

Piece 1 will be the entire rectangle, 12 inch $\times$ 14 inch. Piece 2 will be the two negative sections, which are $4\frac{1}{2}$ inch $\times$ 12 inch each (Figure 6-15).

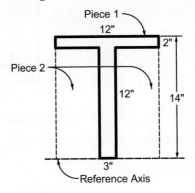

FIGURE 6-15

TABLE 6-7

Piece	Area (in²)	y (in)	Ay (in³)
1	168	7	1,176
2	−108	6	−648
Σ	60		528

$$\bar{y} = \frac{\Sigma Ay}{\Sigma A} = \frac{528 \text{ in}^3}{60 \text{ in}^2} = 8.8 \text{ in}$$

This verifies our previous calculation that the location of the neutral axis of the T section is 8.8 inches up from the bottom of the section. Now we can calculate the moment of inertia, I_x, but first we have to calculate the moment of inertia for the pieces.

$$I_{c1} = \frac{bh^3}{12} = \frac{12 \text{ in}(14 \text{ in})^3}{12} = 2,744 \text{ in}^4$$

$$I_{c2} = \frac{bh^3}{12} = -\frac{9 \text{ in}(12 \text{ in})^3}{12} = -1,296 \text{ in}^4$$

TABLE 6-8

Piece	Area (in²)	y (in)	Ay (in³)	I_c (in⁴)	d_y (in)	d_y^2 (in²)	Ad_y^2 (in⁴)	$I_{cx} + Ad_y^2$ (in⁴)
1	168	7	1,176	2,744	1.8	3.25	544.3	3,288.3
2	−108	6	−648	−1,296	2.8	7.8	−846.7	−2,142.7
Σ	60		528					1,145.6

Let's take another look at the angle in Figure 6-8 but this time we will flip the angle horizontally and determine which axis is stiffer and stronger (Figure 6-16).

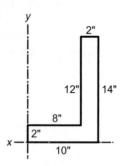

FIGURE 6-16

We will break the angle into slightly different pieces than we did previously (Figure 6-17).

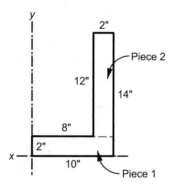

FIGURE 6-17

We will begin by finding the moment of inertia about the x axis.

$$I_{cx1} = \frac{bh^3}{12} = \frac{10 \text{ in}(2 \text{ in})^3}{12} = 6.7 \text{ in}^4$$

$$I_{cx2} = \frac{bh^3}{12} = \frac{2 \text{ in}(12 \text{ in})^3}{12} = 288 \text{ in}^4$$

TABLE 6-9

					I_x				
Piece	Area (in²)	y (in)	Ay (in³)	I_{cx} (in⁴)	d_y (in)	d_y^2 (in²)	Ad_y^2 (in⁴)	$I_{cx} + Ad_y^2$ (in⁴)
1	20	1	20	6.7	3.8	14.4	288.8	295.5
2	24	8	192	288	3.2	10.2	245.8	533.8
Σ	44		212					829.6

$$\bar{y} = \frac{\Sigma Ay}{\Sigma A} = \frac{212 \text{ in}^3}{44 \text{ in}^2} = 4.8 \text{ in}$$

Having I_x we can now calculate the moment of inertia about the y axis.

$$I_{cy1} = \frac{bh^3}{12} = \frac{2 \text{ in}(10 \text{ in})^3}{12} = 166.7 \text{ in}^4$$

$$I_{cy2} = \frac{bh^3}{12} = \frac{12 \text{ in}(2 \text{ in})^3}{12} = 8.0 \text{ in}^4$$

$$\bar{x} = \frac{\Sigma Ax}{\Sigma A} = \frac{316 \text{ in}^3}{44 \text{ in}^2} = 7.2 \text{ in}$$

TABLE 6-10

				I_y				
Piece	Area (in²)	x (in)	Ax (in³)	I_{cy} (in⁴)	d_x (in)	d_x^2 (in²)	Ad_x^2 (in⁴)	$I_c + Ad_x^2$ (in⁴)
1	20	5	100	166.7	2.2	4.8	95.2	261.9
2	24	9	216	8	1.8	3.3	79.3	87.3
Σ	44		316					349.2

A comparison of the moments of inertia shows $I_x(829.6 \text{ in}^4)$ is stronger than $I_y(349.2 \text{ in}^4)$. For this section we can say that the x axis is the *strong axis* and y axis is the *weak axis*. These are terms we will find commonly applied to structural sections.

Our next challenge is to determine how to use the centroid and moment of inertia in the analysis and design of structural members. We will cover this issue in Chapter 7.

Sample Problems

Sample Problem 6-1: Determine $\bar{y}$ and I_x for the composite section of the structural member shown in Figure S6-1 on page 168.

SOLUTION

Step 1   The section shown in Figure S6-1 is symmetrical about the y axis; therefore, we only have to be concerned about the x axis. We can combine the two outside pieces, which are identical, when calculating $\bar{y}$ and I_x (Figure S6-2).

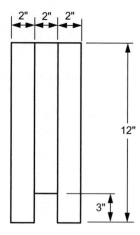

FIGURE S6-1

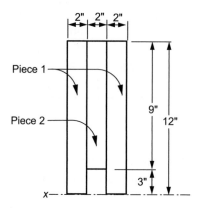

FIGURE S6-2

Step 2 Calculate the moment of inertia of the pieces about the x axis.

$$I_{cx1} = 2\left(\frac{bh^3}{12}\right) = 2\left(\frac{2\text{ in}(12\text{ in})^3}{12}\right) = 576\text{ in}^4$$

$$I_{cx2} = \frac{bh^3}{12} = \frac{2\text{ in}(9\text{ in})^3}{12} = 121.5\text{ in}^4$$

Step 3 Having the moments of inertia of the components, we can proceed to construct our table (Table S6-1).

TABLE S6-1							
Piece	Area (in^2)	y (in)	Ay (in^3)	I_c (in^4)	d_y (in)	d_y^2 (in^2)	Ad_y^2 (in^4)	$I_{cx} + Ad_y^2$ (in^4)
1	48	6	288	576	.4	.16	7.7	583.7
2	18	7.5	135	121.5	1.1	1.2	21.8	143.3
Σ	66		423					727

$$\bar{y} = \frac{\Sigma Ay}{\Sigma A} = \frac{423 \text{ in}^3}{66 \text{ in}^2} = 6.4 \text{ in}$$

Step 4 It is not always possible, but in this case we can verify our answer by using the positive moment of inertia plus negative moment of inertia (Figure S6-3).

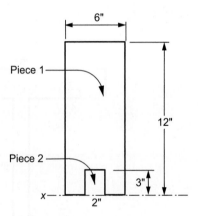

FIGURE S6-3

Step 5 Calculate the moment of inertia of the components about the x axis.

$$I_{cx1} = \frac{bh^3}{12} = \frac{6 \text{ in}(12 \text{ in})^3}{12} = 864 \text{ in}^4$$

$$I_{cx2} = -\frac{bh^3}{12} = -\frac{2 \text{ in}(3 \text{ in})^3}{12} = -4.5 \text{ in}^4$$

TABLE S6-2

Piece	Area (in^2)	y (in)	Ay (in^3)	I_c (in^4)	d_y (in)	d_y^2 (in^2)	Ad_y^2 (in^4)	$I_{cx} + Ad_y^2$ (in^4)
1	72	6	432	86.4	.4	.17	12.1	876
2	-6	1.5	-9	-4.5	4.9	24.1	-144.6	-149
Σ	66		423					727

$$\bar{y} = \frac{\Sigma Ay}{\Sigma A} = \frac{423 \text{ in}^3}{66 \text{ in}^2} = 6.4 \text{ in}$$

Result

The moment of inertia about the x axis is 727 in^4 and is verified in step 5.

Sample Problem 6-2: Determine $\bar{y}$ and L_x for the composite section of the structural member shown in Figure S6-4. Once again, the section is symmetrical about the y axis and therefore we only have to calculate the moment of inertia about the x axis (Figure S6-4).

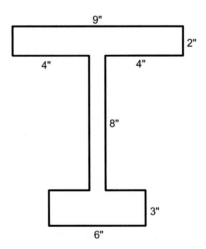

FIGURE S6-4

SOLUTION

Step 1 Break the composite section into pieces (Figure S6-5).

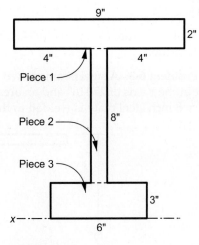

FIGURE S6-5

Step 2   Determine the moment of inertia for the pieces of the section.

$$I_{cx1} = \frac{bh^3}{12} = \frac{9 \text{ in}(2 \text{ in})^3}{12} = 6.0 \text{ in}^4$$

$$I_{cx2} = \frac{bh^3}{12} = \frac{1 \text{ in}(8 \text{ in})^3}{12} = 42.7 \text{ in}^4$$

$$I_{cx3} = \frac{bh^3}{12} = \frac{6 \text{ in}(3 \text{ in})^3}{12} = 13.5 \text{ in}^4$$

Step 3   Having the moments of inertia of the pieces, we can proceed to construct our table (Table S6-3).

TABLE S6-3								
Piece	Area (in^2)	y (in)	Ay (in^3)	I_c (in^4)	d_y (in)	d_y^2 (in^2)	Ad_y^2 (in^4)	$I_{cx} + Ad_y^2$ (in^4)
1	18	12	216	6	5.2	27	486.7	492.7
2	8	7	56	42.7	.2	.04	.32	43
3	18	1.5	27	13.5	5.3	28.1	505.6	519.1
Σ	44		299					1,054.8

$$\bar{y} = \frac{\Sigma Ay}{\Sigma A} = \frac{299 \text{ in}^3}{44 \text{ in}^2} = 6.8 \text{ in}$$

Results

The moment of Inertia, L_x, equals 1,054.8 in^4, which is a measure of stiffness of the composite section.

Sample Problem 6-3: A steel beam shown in Figure S6-6 above has a moment of inertia abut the x axis of 272 in^4 and an area of 14.4 in^2. What is the increase in l_x if a 1 inch $\times$ 8 inch steel plate is welded to the bottom of the beam?

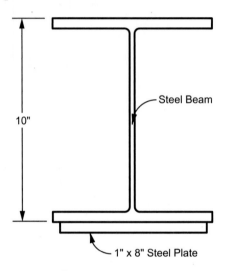

10"

Steel Beam

1" x 8" Steel Plate

FIGURE S6-6

SOLUTION

Step 1 Determine the moment of inertia, l_x, of the steel plate.

$$l_x = \frac{bh^3}{12} = \frac{8 \text{ in } (1 \text{ in})^3}{12} = .67 \text{ in}^4$$

Step 2 Using the bottom of the steel plate as our reference, we can calculate I_x of the composite section.

TABLE S6-4

Piece	Area (in^2)	y (in)	Ay (in^3)	I_c (in^4)	d_y (in)	d_y^2 (in^2)	Ad_y^2 (in^4)	$I_{cx} + Ad_y^2$ (in^4)
Plate	8	.5	4	.67	3.54	12.5	100	100.7
Beam	14.4	6	86.4	272	1.96	3.84	55.3	327.3
Σ	22.4		90.4					428.0

$$\bar{y} = \frac{\Sigma Ay}{\Sigma A} = \frac{90.4 \text{ in}^3}{22.4 \text{ in}^2} = 4.04 \text{ in}$$

Result

The initial moment of inertia of the beam about the x axis was 272 in⁴, and with
the plate it increased to 428 in⁴; the increase is $428 - 272$ in⁴ or 156 in⁴.

Sample Problem 6-4: Determine the centroid and moment of inertia about the
x axis and y axis for the $\frac{1}{4}$"-thick metal trough shown in Figure S6-7.

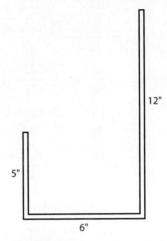

FIGURE S6-7

SOLUTION:

Step 1 Since we need to find the centroid and moment of inertia about both axes,
we will place the trough in a coordinate system indicating the piece numbers (Figure S6-8).

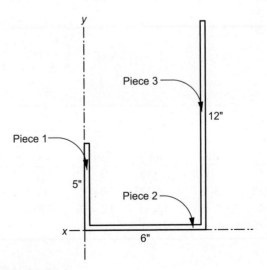

FIGURE S6-8

Step 2 Determine the moment of inertia about both x axis and y axis for the components.

$$I_{xc1} = \frac{bh^3}{12} = \frac{.25 \text{ in}(5 \text{ in})^3}{12} = 2.6 \text{ in}^4$$

$$I_{xc2} = \frac{bh^3}{12} = \frac{5.5 \text{ in}(.25 \text{ in})^3}{12} = .0072 \text{ in}^4 \text{ (Neglect, too small)}$$

$$I_{xc1} = \frac{bh^3}{12} = \frac{.25 \text{ in}(12 \text{ in})^3}{12} = 36 \text{ in}^4$$

$$I_{yc1} = \frac{bh^3}{12} = \frac{5 \text{ in}(.25 \text{ in})^3}{12} = .0065 \text{ in}^4 \text{ (Neglect, too small)}$$

$$I_{yc2} = \frac{bh^3}{12} = \frac{.25 \text{ in}(5.5 \text{ in})^3}{12} = 3.47 \text{ in}^4$$

$$I_{yc3} = \frac{bh^3}{12} = \frac{12 \text{ in}(.25 \text{ in})^3}{12} = .0015 \text{ in}^4 \text{ (Neglect, too small)}$$

Step 3 Using our tables (Table S6-5 and S6-6), I_x and I_y.

TABLE S6-5

					I_x			
Piece	Area (in^2)	y (in)	Ay (in^3)	I_{cx} (in^4)	d_y (in)	d_y^2 (in^2)	Ad_y^2 (in^4)	$I_{cx} + Ad_y^2$ (in^4)
1	1.25	2.5	3.125	2.6	1.19	1.42	1.78	4.38
2	1.375	.125	.17	0	3.57	12.73	17.5	17.5
3	3.00	6	18	36	2.31	5.32	15.97	51.97
Σ	5.625		20.77					73.85

$$\bar{y} = \frac{\Sigma Ay}{\Sigma A} = \frac{20.77 \text{ in}^3}{5.63 \text{ in}^2} = 3.69 \text{ in}$$

TABLE S6-6

					I_y			
Piece	Area (in^2)	x (in)	Ax (in^3)	I_{cy} (in^4)	d_x (in)	d_x^2 (in^2)	Ad_x^2 (in^4)	$I_{cx} + Ad_x^2$ (in^4)
1	1.25	.125	.16	0	3.78	14.25	17.81	17.81
2	1.375	3	4.125	3.47	.9	.81	1.11	4.58
3	3.00	5.88	17.63	0	1.98	3.9	11.7	11.7
Σ	5.625		21.91					33.79

$$\bar{x} = \frac{\Sigma Ax}{\Sigma A} = \frac{21.91 \text{ in}^3}{5.63 \text{ in}^2} = 3.9 \text{ in}$$

Results

The trough has an I_x of 73.85 in^4 and I_y of 33.79 in^2. The centroid $\bar{y}$ and $\bar{x}$ are shown in Figure S6-9.

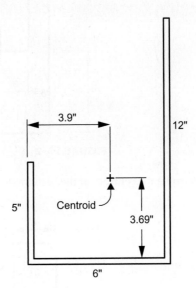

FIGURE S6-9

Supplementary Exercises

6-1. Determine the centroid of the section shown in Figure E6-1.

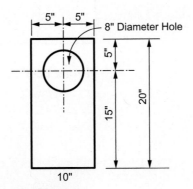

FIGURE E6-1

6-2. Determine the centroid of the section shown in Figure E6-2 (SI units).

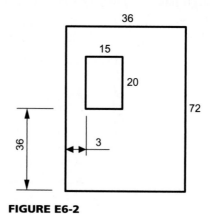

FIGURE E6-2

6-3. Determine the centroid of the section shown in Figure E6-3.

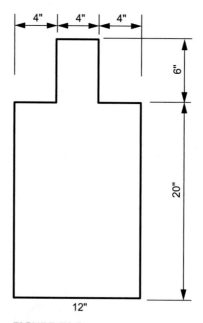

FIGURE E6-3

6-4. Determine the centroid of the section shown in Figure E6-4.

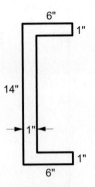

FIGURE E6-4

6-5. Determine the moment of inertia of the section shown in Figure E6-4.
6-6. Determine the moment of inertia of the section shown in Figure E6-5 (SI units).

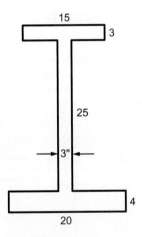

FIGURE E6-5

6-7. The steel beam shown in Figure E6-6 has a moment of inertia about the x axis of 103 in^4 and an area of 4.71 in^2. If a $\frac{1}{4}$ " $\times$ 6 inch steel plate is welded to the bottom of the beam, what in the moment of inertia of the composite section?

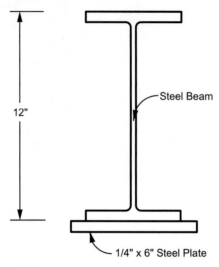

FIGURE E6-6

6-8. The steel beam shown in Figure E6-7 has a moment of inertia about the x axis of 11,493 cm^4 and an area of 94.8 cm^2. Determine l_x of the composite section if a 1×15 cm plate is welded to the top of the beam and a 1×10 cm plate is welded to the bottom of the beam.

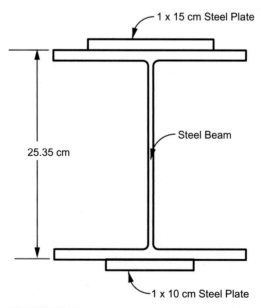

FIGURE E6-7

CHAPTER 7

Beam Analysis and Design

7.1 BENDING STRESSES

In Chapter 4 we learned that, when our friendly gorilla stands on a simply supported beam, he introduces compression in the top of the beam and tension in the bottom of the beam (Figure 4-4). In Chapter 5 we determined that the reason the beam did not collapse under the force of the gorilla's weight was due to an internal shear and resisting moment, M_R. Now we can explore how these two issues are directly related.

If the beam the gorilla is standing on has a rectangular cross section, we know the centroid or neutral axis is located at mid-height (Chapter 5), as shown in Figure 7-1.

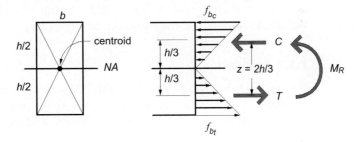

FIGURE 7-1

Bending stresses that are within the elastic limit (Chapter 4) are composed of compressive and tensile stress as shown in Figure 7-1. The compressive bending stresses, (f_{bc}), and tension stresses, (f_{bt}), are distributed over the areas $b \times h/2$, and the resultants of these stresses, C and T, are equal to $1/2 f(h/2)(b)$. The resultants act at the centroid of the triangle, which is located at $2/3(h/2)$ or $h/3$ up from the neutral axis. C and T form a couple and the distance between them (which we will call z) must be $2 \times h/3$ or $2h/3$. To determine the internal resisting moment,

M_R, we recall from Chapter 2, we multiply one of the forces times the distance between them.

$$M_R = Cz = Tz$$

$$= C\left(\frac{2h}{3}\right) = T\left(\frac{2h}{3}\right)$$

$$= \frac{1}{2}f_b\left(\frac{h}{2}\right)b\left(\frac{2h}{3}\right)$$

$$= f_b\left(\frac{bh^2}{6}\right)$$

This new term derived above, $bh^2/6$, is called the *section modulus* and is directly related to the moment of inertia.

If we use the term c to define the distance from the neutral axis to the location of the highest stresses, which are commonly called the beams *extreme fibers*, we can clarify this relationship between the section modulus and the moment of inertia. If the beam is rectangular or symmetrical about the neutral axis, c equals $h/2$.

$$M_R = f_B \frac{I}{c} = f_B\left(\frac{bh^3}{12}\right)\frac{2}{h} = f_B\left(\frac{bh^2}{6}\right)$$

Using the moment of inertia, we arrive at the same solution as we did above using the stress to derive the section modulus.

If we want to find the internal resisting moment, (M_R), of a beam that has an asymmetrical cross section, we would no longer be able to use the term $c = h/2$. We would have to use the equation: $M_R = f_b(I/c)$.

7.1.1 Analysis

We can begin by using these equations, $M_R = bh^3/6$ and $M_R = f_b(I/c)$, in analysis problems. In this analysis problem, such as the one shown in Figure 7-2, we are given a member with a cross section, its size and loads, and must find the actual bending stress.

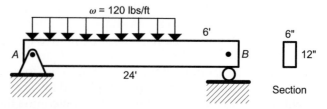

ω = 120 lbs/ft

6'

6"

A

B

12"

24'

Section

FIGURE 7-2

To determine the internal resisting moment the beam must support, we have to calculate the reactions and draw the shear and moment diagrams (Figure 7-3).

$$W = \omega L_L = \frac{120 \text{ lbs}}{\text{ft}}(18 \text{ ft}) = 2{,}160 \text{ lbs} \ (@9 \text{ ft from } R_{Ay})$$

$$\Sigma M_A = 0 = -2{,}160 \text{ lbs}(9 \text{ ft}) + R_B(24 \text{ ft})$$

$$R_B = 810 \text{ lbs}$$

$$\Sigma F_y = -2{,}160 \text{ lbs} + 810 \text{ lbs} + R_{Ay}$$

$$R_{Ay} = 1{,}350 \text{ lbs}$$

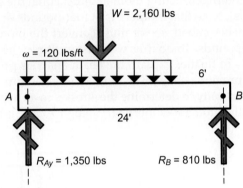

FIGURE 7-3

Following the procedure we established in Chapter 5, we draw the shear and moment diagrams (Figure 7-4).

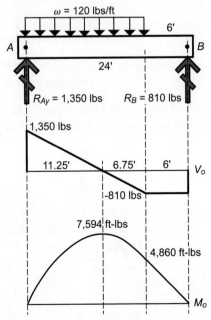

FIGURE 7-4

Having the maximum moment, we can analyze the beam's cross section to determine the section modulus and actual stress under these loading conditions. From Figure 7-2 we have a b of 6 inches and an h of 12 inches.

$$S = \frac{bh^2}{6} = \frac{6 \text{ in}(12 \text{ in})^2}{12} = 72 \text{ in}^3$$

$$f_b = \frac{M}{S} = \frac{7,594 \text{ ft-lbs}(12 \text{ in/ft})}{72 \text{ in}^3} = 1,266 \text{ lbs/in}^2 \text{ or } 1,266 \text{ psi}$$

In the last equation we use M rather than M_R (this is just a simplification because the internal resisting moment must equal the external moment due to the applied loads). Note that moment is in foot-pounds or kip-feet and the section modulus is in inches cubed, so we must convert the moment to the units of inch-pounds or kip-pounds. To do this, we must multiply the moment by 12 inches per foot.

To further explain the beam analysis process, we will use Figure 6-14 from the previous chapter shown below as Figure 7-5. This will provide us with the opportunity to determine the stresses in an asymmetrical cross section, and we can use the moment of inertia (I) value we calculated in the Chapter 6: $I = 1,145.6 \text{ in}^4$.

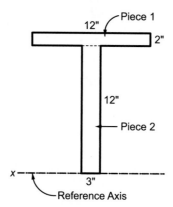

FIGURE 7-5

We will assume the beam is simply supported and must support the loads shown in Figure 7-6, and the weight of the beam we will assume to be 18 pounds per foot.

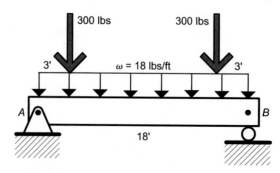

FIGURE 7-6

We need the load diagram so we will begin by calculating the reactions.

$$W = \omega L_L = 18 \text{ lb/ft}(18 \text{ ft}) = 324 \text{ lbs}$$

$$\Sigma M_A = 0$$

$$0 = -300 \text{ lbs}(3 \text{ ft}) - 300 \text{ lbs}(15 \text{ ft}) - 324 \text{ lbs}(9 \text{ ft}) + R_B(18 \text{ ft})$$

$$R_B = 462 \text{ lbs}$$

$$\Sigma F_y = 0$$

$$0 = -300 \text{ lbs} - 300 \text{ lbs} - 324 \text{ lbs} + 462 \text{ lbs} + R_{Ay}$$

$$R_{Ay} = 462 \text{ lbs}$$

Having the reactions we can now construct the load, shear, and moment diagram (Figure 7-7).

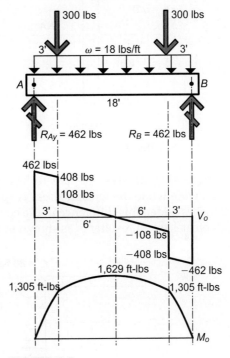

FIGURE 7-7

We calculated the maximum moment of 1,629 foot pounds. From this we can determine the maximum bending stress in the beam. Since the beam cross section is asymmetrical, we will need two c values, one for compression in the top (c_c) and one for tension in the bottom (c_t). From these values we will have to calculate both the designate bending stresses for compressive bending stress f_{bc} and for tensile bending stresses f_{bt}. We will begin by calculating c_c and c_t.

$$c_t = \bar{y} = 8.8 \text{ in}$$

$$c_c = h - \bar{y} = 14 \text{ in} - 8.8 \text{ in} = 5.2 \text{ in}$$

Having the c values, we can calculate the compression and tension stress in the section.

$$f_{b_t} = \frac{M c_t}{I} = \frac{1{,}629 \text{ ft-lbs}(12 \text{ in/ft})(8.8 \text{ in})}{1{,}145.6 \text{ in}^4} = 150.2 \text{ lbs/in}^2$$

$$f_{b_c} = \frac{M c_c}{I} = \frac{1{,}629 \text{ ft-lbs}(12 \text{ in/ft})(5.2 \text{ in})}{1{,}145.6 \text{ in}^4} = 88.7 \text{ lbs/in}^2$$

We have to remember that T and C form a couple and must have equal magnitude. This accounts for the difference in f_{bc} and f_{bt}. If we examine each stress block, we can see how these stresses produce equal resultants (Figure 7-8).

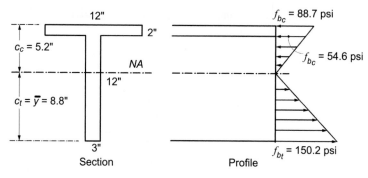

FIGURE 7-8

The compressive stresses 88.7 psi to 54.6 psi are spread over an area that is 12 inches wide while the tension stress is spread continuously over a area that is only 2 inches wide. This difference in the configuration of area and distribution of stresses creates equal resultants as shown in Figure 7-9.

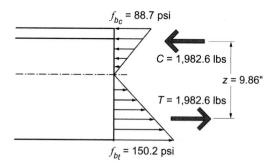

FIGURE 7-9

In addition to problems such as the one above, where we are given the properties of the member and must find the actual bending stresses, there are other types of analysis problems. In the following example we are asked to determine the load carrying capacity of a beam given its properties. In this problem we will determine the magnitude uniform distributed load that a 4 inch × 12 inch beam can support if it spans 16 feet and the allowable bending stress (F_b) is 1,450 psi.

Since we do not have the magnitude of the uniform distributed load, we will have to draw the load, shear, and moment diagrams in terms of ω (Figure 7-10).

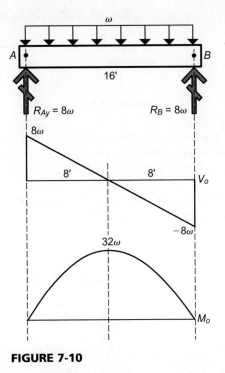

FIGURE 7-10

Having the moment, we can proceed with the determination of ω.

$$F_b = \frac{M}{S}$$

$$M = F_b S = F_b\left(\frac{bh^2}{6}\right)$$

$$32\omega = \left(\frac{1,450 \text{ lbs}}{\text{in}^2}\right)\left(\frac{4 \text{ in}(12 \text{ in})^2}{6}\right)$$

$$\omega = 4,500 \text{ lbs/in} = 362.5 \text{ lbs/ft}$$

7.1.2 Design

In many ways design of structural members for moment is very similar to analysis. When designing beams we know some of the characteristics such as loads and span and must calculate the minimum size beam to support these conditions.

In the next problem let's assume that we know the beam has a 4-inch wide rectangular cross section and we must determine the most economical depth of beam to support the loads in bending (Figure 7-11). For this problem we will assume an allowable bending stress of 1,200 psi.

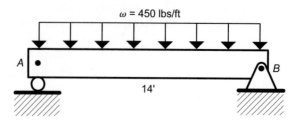

FIGURE 7-11

The beam supports a uniform distributed load of 450 lbs/ft, which includes its own weight. The beam load is symmetrical, so we can assume the reactions are equal to 7 ft × 450 lbs/ft or 3,150 lbs.

Having the reactions, we can draw the load, shear, and moment diagram (Figure 7-12).

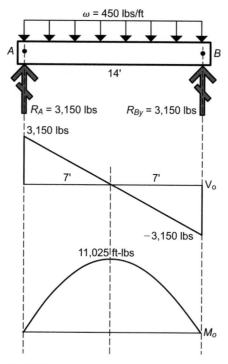

FIGURE 7-12

With the maximum moment of 11,025 ft-lbs, we can calculate the required minimum beam depth.

$$F_b = \frac{M}{S}$$

$$S = \frac{M}{F_b}$$

$$\frac{bh^2}{6} = \frac{(4 \text{ in})h^2}{6} = \frac{M}{F_b}$$

$$h^2 = \frac{(6)(11{,}025 \text{ ft-lbs})(12 \text{ in/ft})}{4 \text{ in}(1{,}200 \text{ lbs/in}^2)}$$

$$h^2 = 165.4 \text{ in}^2$$

$$h = 12.9 \text{ in}$$

If we want to keep the beam dimensions in full inches, we would use a 4 inch × 13 inch beam.

7.2 SHEAR STRESSES

In Chapter 4 we learned a simple explanation of *shear stress*, which is the resistance of one plane of material to sliding over the adjacent plane. Shear stresses, unlike bending stresses, are both horizontal and vertical.

Once again we will call on our friendly gorilla to demonstrate how shear stresses are distributed in the beam (Figure 7-13).

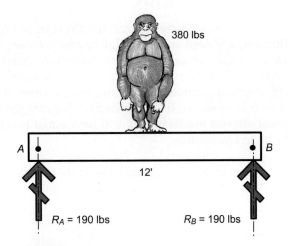

FIGURE 7-13

We will assume the gorilla is standing at the center of the beam, which is in static equilibrium. Since the gorilla is standing in the center of the beam, we can calculate the reactions as half of his weight or 190 pounds each. Suppose we take a section at 7 feet (Figure 7-14 shown on page 188).

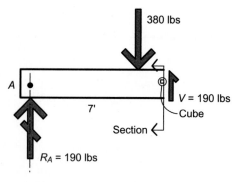

FIGURE 7-14

As we learned in Chapter 4, the shear stress is distributed over the face of the section. At 7 feet, if we will extract and enlarge the cube shown on the face of the cross section in Figure 7-14, we can observe and calculate f_v.

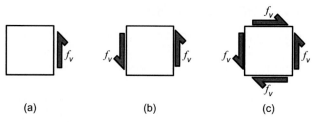

FIGURE 7-15

It is immediately evident that the cube in Figure 7-15(a) is not in static equilibrium. We need an equal and opposite force on the other side of the cube to maintain the sum of the forces in the vertical direction equal to zero. If we have the vertical forces balanced as shown in (b), the cube will rotate in a counterclockwise direction. To stabilize the cube, we need another shear couple acting in a clockwise direction as shown in (c). From this diagram we can see that if we have vertical shear stresses in a beam, we will have equal horizontal shear stresses to keep the beam in equilibrium.

To calculate the distribution of shear stresses over the face of the cross section, we will need the general shear formula:

$$f_v = \frac{QV}{Ib}$$

where:

Q is the moment of area above the shear plane under consideration about the neutral axis,

V is the shear obtained from the shear diagram,

I is the moment of inertia, and

b is the width of the member at the shear plane.

To understand the distribution of shear stresses, let's take a 4 inch × 8 inch rectangular beam with a shear *(V)* of 190 pounds acting on its cross section. For illustration purposes we will begin by investigating the shear stress at 1 inch below the top of the beam and continue at 1 inch intervals down to the neutral axis (Figure 7-16).

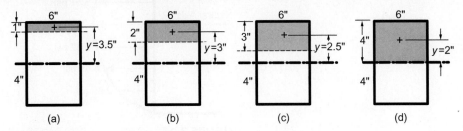

FIGURE 7-16

Using the general shear formula we can calculate the shear stress in each of the sections shown in Figure 7-16.

$$I = \frac{bh^3}{12} = \frac{6 \text{ in}(8 \text{ in})^3}{12} = 256 \text{ in}^4$$

Section (a)

$$Q = Ay = (1 \text{ in})(6 \text{ in})(3.5 \text{ in}) = 21 \text{ in}^3$$

$$f_v = \frac{QV}{Ib} = \frac{(21 \text{ in}^3)(190 \text{ lbs})}{(256 \text{ in}^4)(6 \text{ in})} = 2.60 \text{ psi}$$

Section (b)

$$Q = Ay = (2 \text{ in})(6 \text{ in})(3 \text{ in}) = 36 \text{ in}^3$$

$$f_v = \frac{QV}{Ib} = \frac{(36 \text{ in}^3)(190 \text{ lbs})}{(256 \text{ in}^4)(6 \text{ in})} = 4.45 \text{ psi}$$

Section (c)

$$Q = Ay = (3 \text{ in})(6 \text{ in})(2.5 \text{ in}) = 45 \text{ in}^3$$

$$f_v = \frac{QV}{Ib} = \frac{(45 \text{ in}^3)(190 \text{ lbs})}{(256 \text{ in}^4)(6 \text{ in})} = 5.57 \text{ psi}$$

Section (d)

$$Q = Ay = (4 \text{ in})(6 \text{ in})(2 \text{ in}) = 48 \text{ in}^3$$

$$f_v = \frac{QV}{Ib} = \frac{(48 \text{ in}^3)(190 \text{ lbs})}{(256 \text{ in}^4)(6 \text{ in})} = 5.94 \text{ psi}$$

Since the shear stresses will be mirrored below the neutral axis, we can now plot the stress distribution (Figure 7-17).

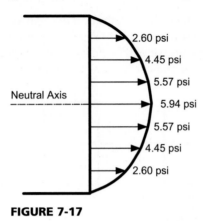

2.60 psi

4.45 psi

5.57 psi

Neutral Axis

5.94 psi

5.57 psi

4.45 psi

2.60 psi

FIGURE 7-17

From this plot we see that shear stresses are maximum at the neutral axis, and since there is no area above the extreme fibers, the shear stresses at the top and bottom of the beam are zero.

The general shear formula can be used on beams of any cross section, but the formula can be simplified for beams with a rectangular (or square) cross section. Taking the formula for the moment of inertia of a beam rectangular cross section and substituting this value into the general shear formula, we create a formula that can be used to determine the maximum shear stress (at the neutral axis) in beams with a rectangular cross section.

$$Q = Ay = \frac{h}{2}(b)\frac{h}{4} = \frac{bh^2}{8}$$

$$I = \frac{bh^3}{12}$$

$$f_v = \frac{QV}{Ib} = \frac{bh^2}{8} \times \frac{12}{bh^3} \times \frac{V}{b}$$

$$= \frac{3V}{2bh} = \frac{3V}{2A}$$

Note:

This formula can only be used for beams with a rectangular or square cross section.

7.3 DEFLECTION

As previously noted, all beams are designed for moment, shear, and deflection. In this section we will discuss deflection, which can be simply defined as vertical sag. If a beam lacks adequate stiffness, it will appear to be unsafe or, if supporting a floor, create excessive bounce. More importantly, excessive deflection can cause structural failure. For example, a flat or nearly flat roof beam with excessive deflection will sag under the weight of water collected from rain or snow, causing a "ponding" effect, which may lead to a total roof collapse.

The deflection limits are established by code and are expressed in terms of span. For roofs the total load deflection is normally limited to the span divided by 180 or $\ell/180$. Roof live load deflection is limited to $\ell/240$. For floors the deflection is limited to the $\ell/240$ and the live load limited to $\ell/360$. These limitations are denoted by the symbol $\Delta_{allowable}$. The calculated deflection will be denoted as Δ_{actual}.

Deflection formulas are organized in the following format:

$$\Delta_{actual} = \text{a constant} \times W \text{ or } P\left(\frac{L^3}{EI}\right)$$

where: the constant is related to the type of load and support conditions,

W (ωL) or P the total load types,

L the span of the beam in feet,

E the modulus of elasticity, and

I the moment of inertia.

We can see how this format applies by examining the deflection equation for a beam supporting a uniform distributed load.

$$\Delta_{actual} = \frac{5\,WL^3}{384\,EI} = \frac{5\,\omega L^4}{384\,EI}$$

In this equation 5/384 is the constant for uniform distributed load on a simply supported beam, W is the total load, and EI a measure of the stiffness of the beam due to the materials' modulus of elasticity (E) and the beam's moment of inertia (I).

To avoid making mistakes in the units (feet, psi, etc.), we multiply all deflection equations in U.S. units by a conversion factor (1,728 in^3/ft^3), thus eliminating the need to convert feet to inches, foot-pounds into inch-pounds or pounds per inch squared into pounds per foot squared. In SI units, we use 10^3 mm/m.

We can simplify our beam calculations by using formulas available in a number of design manuals and handbooks. A few commonly used formulas for moment, shear, and deflection are given in Table 7-1 on pages 192–193.

It is extremely important to realize that these formulas apply **only** to the loading conditions shown. We can, however, add formulas from the table. If we have a beam with a uniform distributed load, Table 7-1(a), and two equally spaced concentrated loads, Table 7-1(c), we can add the moment, shear, and deflection formulas.

If we have a problem that presents a loading condition that is not shown in Table 7-1, we must draw shear and bending diagrams. In these cases we can

TABLE 7-1

(a)	Uniform Distributed Load—Simple Span

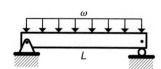

$$M = \frac{\omega L^2}{8}$$

$$V = R = \frac{wL}{2}$$

$$\Delta_{actual} = \frac{5\,wL^4(1{,}728)}{384\,EI}$$

(b)	Concentrated Load at Mid-Span—Simple Span

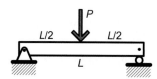

$$M = \frac{PL}{4}$$

$$V = \frac{P}{2}$$

$$\Delta_{actual} = \frac{PL^3(1{,}728)}{48\,EI}$$

(c)	Two Equal Concentrated Loads Symmetrically Placed—Simple Span

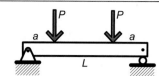

$$M = Pa$$

$$V = R = P$$

$$\Delta_{actual} = \frac{Pa(3L^2 - 4a^2)(1{,}728)}{24\,EI}$$

(d)	Linear Distributed Loads—Simple Span

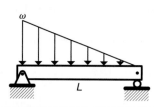

$$M = \frac{\omega L^2}{9\sqrt{3}}$$

$$V = R_{left} = \frac{2\omega L}{3}$$

$$\Delta_{actual} = \frac{.0065\omega L^4(1{,}728)}{EI}$$

(e)	Two Linear Distributed Loads—Simple Span

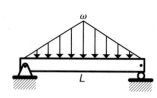

$$M = \frac{\omega L^2}{12}$$

$$V = R = \frac{\omega L}{4}$$

$$\Delta_{actual} = \frac{\omega L^4(1{,}728)}{120\,EI}$$

(continued)

TABLE 7-1   (CONTINUED)

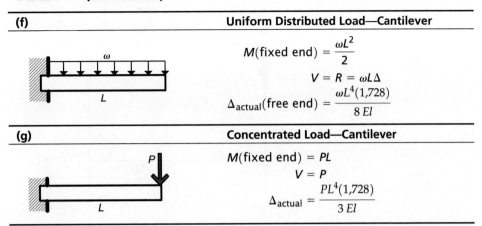

(f)	Uniform Distributed Load—Cantilever
	$M(\text{fixed end}) = \dfrac{\omega L^2}{2}$
	$V = R = \omega L \Delta$
	$\Delta_{actual}(\text{free end}) = \dfrac{\omega L^4(1{,}728)}{8\,EI}$
(g)	Concentrated Load—Cantilever
	$M(\text{fixed end}) = PL$
	$V = P$
	$\Delta_{actual} = \dfrac{PL^4(1{,}728)}{3\,EI}$

approximate the deflection. We will begin by approximating the deflection for a beam with an overhang. In this case the formula in Table 7-1(a) can be used to approximate deflection. The overhang will relieve some of the moment, and consequently deflection, in the span between supports so this method will provide us with a conservative calculation of the actual deflection.

To demonstrate this method of approximate deflection, we will use the beam with an overhang shown in Figure 7-18. It assumed modulus of elasticity of 1.5×10^6 psi, and it has a cross section of 4 inches $\times$ 12 inches.

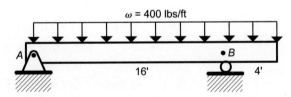

FIGURE 7-18

Before we can determine the deflection, we have to draw the load, shear, and moment. To obtain the loading diagram, we will have to calculate the reactions.

$$W = \omega L_L = 400 \text{ lbs/ft}(20 \text{ ft}) = 8{,}000 \text{ lbs}$$

$$\Sigma M_A = -\frac{WL}{2} + R_B(16 \text{ ft}) = 8{,}000 \text{ lbs}(10 \text{ ft}) + R_B(16 \text{ ft})$$

$$R_B = \frac{80{,}000 \text{ ft-lbs}}{16 \text{ ft}} = 5{,}000 \text{ lbs}$$

$$\Sigma F_y = 0 = -8{,}000 \text{ lbs} + 5{,}000 \text{ lbs} + R_{Ay}$$

$$R_{Ay} = 3{,}000 \text{ lbs}$$

Having the maximum moment from the diagram (Figure 7-19 on the next page), we can proceed to calculate the approximate deflection (between the supports).

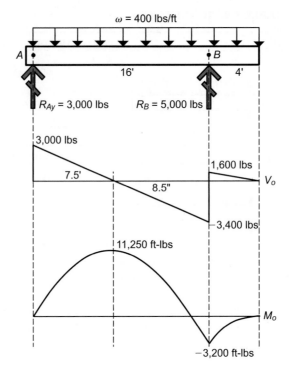

FIGURE 7-19

$$I = \frac{bh^3}{12} = \frac{4 \text{ in}(12 \text{ in})^3}{12} = 576 \text{ in}^4$$

$$\Delta_{\text{Approx.}} = \frac{5 \, \omega L^4 (1{,}728)}{384 \, EI} = \frac{5(400 \text{ lbs/ft})(16 \text{ ft}^4)(1{,}728)}{384(1.5 \times 10^6 \text{ lbs/in}^2)(576 \text{ in}^4)} = .68 \text{ in}$$

If the total load deflection is limited to $L/240$ or .8 inch, these calculations would show that the beam falls within the allowable limits ($\Delta_{\text{approx.}} < \Delta_{\text{allowable}}$).

Note:

This approximate method should only be used with beams that have the length of the overhang equal to, or less than, 1/3 of the spans between the supports.

When a beam is simply supported without an overhang, we can apply another method of approximating the deflection. Using this method we begin by drawing the load, shear, and moment diagram to determine the maximum moment. Once we have the maximum moment, we calculate the uniform distributed load (ω), which would create the same moment. Using the deflection equation for the uniform distributed load, we can approximate the deflection.

To demonstrate this method of approximate deflection, we will use a 6 inch × 8 inch beam that has a moment of inertia I of 256 in^4 and a modulus of elasticity of 1.7 × 10^6 psi (Figure 7-20).

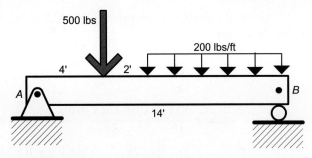

FIGURE 7-20

To obtain the maximum moment, we will calculate the reactions and draw the diagrams.

$$W = \omega L_L = 200 \text{ lbs/ft}(8 \text{ ft}) = 1{,}600 \text{ lbs}$$

$$\Sigma M_A = 0 = -500 \text{ lbs}(4 \text{ ft}) - 1{,}600 \text{ lbs}(10 \text{ ft}) + R_B(14 \text{ ft})$$

$$R_B = \frac{18{,}000 \text{ ft-lbs}}{14 \text{ ft}} = 1{,}285.7 \text{ lbs}$$

$$\Sigma F_y = -500 \text{ lbs} - 1{,}600 \text{ lbs} + 1{,}285.7 \text{ lbs} + R_{Ay}$$

$$R_{Ay} = 814.3 \text{ lbs}$$

Using these reactions we can construct the load, shear, and moment diagrams (Figure 7-21).

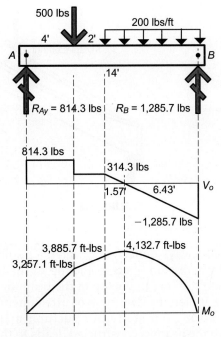

FIGURE 7-21

Using the maximum moment of 4,132.7 ft-lbs, we can calculate the approximate deflection.

$$M = \frac{\omega L^2}{8}$$

$$\omega = \frac{8M}{L^2} = \frac{(8)4{,}132.7 \text{ ft-lbs}}{14 \text{ ft}^2} = 168.7 \text{ lbs/ft}$$

$$\Delta_{\text{approx.}} = \frac{5\omega L^4(1{,}728)}{384 \, EI} = \frac{5(168.7 \text{ lbs/ft})(14 \text{ ft})^4(1{,}728)}{384(1.7 \times 10^6 \text{ psi})(256 \text{ in}^4)} = .34 \text{ in}$$

The allowable deflection for total load is .70 inches, so our approximate deflection is well within the allowable limits. Like all structural calculations, when using approximate deflection the designer must exercise good judgment.

7.4   LATERAL STABILITY

To conclude this chapter we will discuss another type of deflection. Deep beams with narrow widths have a tendency to not only deflect vertically, but sideways as well, which must be accounted for and prevented (Figure 7-22).

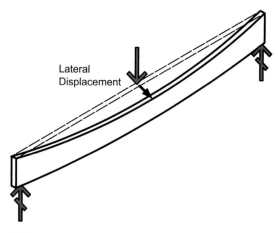

Lateral
Displacement

FIGURE 7-22

This sideways deflection or lateral displacement is caused by the buckling of the compression fibers in the top of the beam. It can be avoided by firmly attaching decking to the top of the beam or by bracing (or bridging) the member along its length. The type of attachment and location of bracing will be addressed in your study of structural materials, as the type of material from which the beam is constructed, leads to the amount of lateral displacement and the type of attachment.

Sample Problems

Sample Problem 7-1: Determine the bending stresses in a beam that has the cross section shown in Figure S7-1 and with a moment of 42 kip-ft.

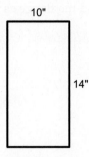

10"

14"

FIGURE S7-1

SOLUTION

Step 1 Since the beam is symmetrical about the x axis, we determine the best way to calculate the moment of inertia is to use the formulas in Table 7-1.

$$I_x = \frac{bh^3}{12} = \frac{10 \text{ in}(14 \text{ in})^3}{12} = 2{,}286.7 \text{ in}^4$$

Step 2 Having the moment of inertia we can now determine the bending stresses in the beam.

$$f_b = \frac{Mc}{I} = \frac{42 \text{ kip-ft}(12 \text{ in/ft})(7 \text{ in})}{2{,}286.7 \text{ in}^4} = 1.54 \text{ kip/in}^2$$

Result

The beam is symmetrical about the x axis; therefore, the bending stresses in compression are equal to the bending stresses in tension.

Sample Problem 7-2: Find the bending stress at the top of the flange (the horizontal portion of the beam) in the inverted T beam shown in Figure S7-2. The beam is simply supported and must resist a moment of 8,000 ft-lbs.

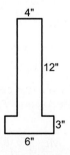

4"

12"

3"

6"

FIGURE S7-2

SOLUTION

Step 1   In order to find the bending stresses we must first determine the location of the centroid or neutral axis and calculate the moment of inertia. We will use the bottom of the beam as the x axis, the web (vertical element) as piece 1 and the flange as piece 2 (Figure S7-3).

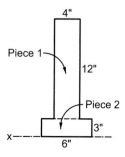

FIGURE S7-3

TABLE S7-1

Piece	Area (in²)	y (in²)	Ay (in³)	I_{cx} (in⁴)	d (in)	d² (in²)	Ad² (in⁴)	$I_c + Ad^2$ (in⁴)
1	48	9	432	576	2.05	4.2	201.7	777.7
2	18	1.5	216	13.5	5.45	29.7	534.7	548.2
Σ	66		459	699.5				1,326

$$\bar{y} = \frac{\Sigma Ay}{\Sigma A} = \frac{459 \text{ in}^3}{66 \text{ in}^2} = 6.95 \text{ in}$$

$$I_{x1} = \frac{bh^3}{12} = \frac{4 \text{ in}(12 \text{ in})^3}{12} = 576 \text{ in}^4$$

$$I_{x2} = \frac{bh^3}{12} = \frac{6 \text{ in}(3 \text{ in})^3}{12} = 13.5 \text{ in}^4$$

Step 2   With this data we can calculate the stress in the bottom or tension side of the cross section using $f_b = Mc/I$.

$$f_b = \frac{Mc}{I} = \frac{8,000 \text{ ft-lbs}(12 \text{ in/ft})(6.95 \text{ in})}{1,326 \text{ in}^4} = 503.2 \text{ lbs/in}^2$$

Step 3 To determine the stress on the top of the flange we can draw the section and use a proportion Figure S7-4.

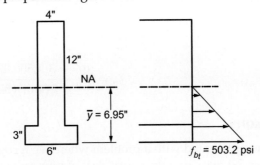

FIGURE S7-4

Step 4 Using similar triangles we can calculate the stress at the top of the flange.

$$\frac{f_{b_t}(\text{top of the flange})}{6.95 \text{ in} - 3 \text{ in}} = \frac{503.2 \text{ lbs/in}^2}{6.95 \text{ in}}$$

$$f_{b_t}(\text{top of the flange}) = 286 \text{ lbs/in}^2$$

Results

By using structural formulas and simple geometry, we can determine the bending stress at any point along the beams cross section.

Sample Problem 7-3: In Figure S7-5 determine the maximum shear and bending stresses in the beam.

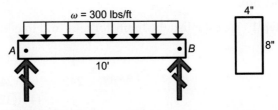

FIGURE S7-5

SOLUTION

Step 1 Begin by using Table 7-1(a) formulas rather than drawing the shear and moment diagrams. We will start with the bending stresses.

$$M = \frac{\omega L^2}{8} = \frac{300 \text{ lbs/ft}(10 \text{ ft})^2}{8} = 3{,}750 \text{ ft-lbs}$$

$$S = \frac{bh^2}{6} = \frac{4 \text{ in}(8 \text{ in})^2}{6} = 42.67 \text{ in}^3$$

$$f_b = \frac{M}{S} = \frac{3{,}750 \text{ ft-lbs}(12 \text{ in/ft})}{42.67 \text{ in}^3} = 1{,}054.7 \text{ psi}$$

Step 2 We now calculate the shear stresses. Once again we can use table formulas and the formula for rectangular cross sections.

$$V = \frac{\omega L}{2} = \frac{300 \text{ lbs}(10 \text{ft})}{2} = 1{,}500 \text{ lbs}$$

$$f_v = \frac{3V}{2A} = \frac{3(1{,}500 \text{ lbs})}{2(4 \text{ in})(8 \text{in})} = 70.3 \text{ psi}$$

Results

When presented with a problem where we recognize the loading conditions from Table 7-1, we can save a great deal of time and effort in reaching a solution.

Sample Problem 7-4: Determine if the deflection in the floor beam shown in Figure S7-6 is within the acceptable limits. Use $E = 29 \times 10^6$ psi and $I_x = 130$ in^4.

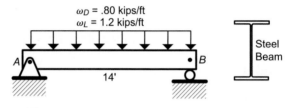

$\omega_D = .80$ kips/ft
$\omega_L = 1.2$ kips/ft

Steel Beam

A   B

14'

FIGURE S7-6

SOLUTION

Step 1 We establish the limits for total and live load.

$$\Delta T_{\text{allowable}} = \frac{\ell}{240} = \frac{14 \text{ ft}(12 \text{ in/ft})}{240} = .70 \text{ in}$$

$$\Delta L_{\text{allowable}} = \frac{\ell}{360} = \frac{14 \text{ ft}(12 \text{ in/ft})}{360} = .50 \text{ in}$$

Step 2 Now that we calculated the limits or allowable deflection, we will use the formulas from Table 7-1 to obtain the actual deflection.

$$\omega_T = \omega_D + \omega_L = .8 \text{ kips/ft} + 1.2 \text{ kips/ft} = 2.0 \text{ kip/ft}$$

$$\Delta T_{\text{actual}} = \frac{5 \, \omega L^4 (1{,}728)}{384 \, EI}$$

$$= \frac{5(2.0 \text{ lbs/ft})(14 \text{ ft})^4 (1{,}728)}{(384)(29 \times 10^6 \text{ lbs/in}^2)(130 \text{ in}^4)}$$

$$= .46 \text{ in} > .70 \text{ in and} > .50 \text{ in}$$

Results

We do not have to calculate the actual live load deflection because Δ_T is less than $\Delta_{L_{\text{Allowable}}}$.

Sample Problem 7-5: Design a beam to support the loads shown in Figure S7-7. The beam must have a cross section in full inches (see Appendix A for Properties of Rectangular Sections). Limit total load deflection to $\ell/240$ and live load deflection to $\ell/360$. Assume a modulus of elasticity of 1.5×10^3 ksi, an allowable bending stress (F_b) of 1,150 psi, and an allowable shear stress (F_v) of 120 psi.

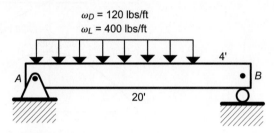

FIGURE S7-7

SOLUTION

Step 1   The loading diagram does not appear in Table 7-1. We will have to draw load, shear, and bending diagrams (Figure S7-8, page 202). We start by calculating the reactions.

$$\omega_T = \omega_D + \omega_L = 120 \text{ lbs/ft} + 400 \text{ lbs/ft} = 520 \text{ lbs/ft}$$

$$W = \omega L_L = 520 \text{ lbs}(16 \text{ ft}) = 8{,}320 \text{ lbs}$$

$$\Sigma M_A = 0 = -8{,}320 \text{ lbs}(8 \text{ ft}) + R_B(20 \text{ ft})$$

$$R_B = 3{,}328 \text{ lbs}$$

$$\Sigma F_y = 0$$

$$= -8{,}320 \text{ lbs} + 3{,}328 \text{ lbs} + R_{Ay}$$

$$R_{AY} = 4{,}992 \text{ lbs}$$

Step 2   Having calculated the maximum moment, we can select a trial size beam based on the section modulus.

$$f_b = \frac{M}{S}$$

$$S = \frac{M}{f_b} = \frac{23{,}962 \text{ ft-lbs}(12 \text{ in/ft})}{1{,}150 \text{ lbs/in}^2} = 250 \text{ in}^3$$

Try: $6'' \times 16'' S = 256 \text{ in}^3$ (see Appendix A)

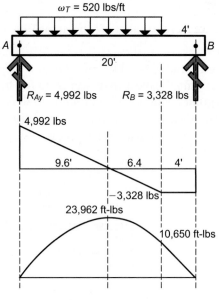

FIGURE S7-8

Step 3 With this trial size we can proceed to check the shear using the highest value from the shear diagram.

$$V = 4{,}992 \text{ lbs} \quad A = 96 \text{ in}^2 \text{ (see Appendix } A\text{)}$$

$$f_v = \frac{3V}{2A} = \frac{3(4{,}992 \text{ lbs})}{2(96 \text{ in}^2)} = 78 \text{ psi} < 120 \text{ psi OK}$$

Step 4 In this last step we need to check deflection. Looking at Table 7-1, we cannot find a similar loading diagram, so we will have to approximate the deflections using the formulas for a uniform distributed load.

$$M = 23{,}962 \text{ ft-lbs}$$

$$w_T = \frac{8M}{L^2} = \frac{8(23{,}962 \text{ ft-lbs})}{(20 \text{ ft})^2} = 479 \text{ lbs/ft}$$

$$\Delta_{T_{\text{allowable}}} = \frac{\ell}{240} = \frac{20 \text{ ft}(12 \text{ in/ft})}{240} = 1.0 \text{ in}$$

$$\Delta_{T_{\text{approx}}} = \frac{5 \, \omega L^4 (1{,}728)}{384 \, EI}$$

$$= \frac{5(479 \text{ lbs/ft})(20 \text{ ft})^4 (1{,}728)}{384(1.5 \times 10^4 \text{ lbs/in}^2)(2{,}731 \text{ in}^4)}$$

$$= .56 \text{ in} > 1.0 \text{ in OK}$$

$$\Delta_{L_{\text{allowable}}} = \frac{\ell}{360} = \frac{20 \text{ ft}(12 \text{ in/ft})}{360} = .67 \text{ in}$$

$$\Delta_{L_{\text{allowable}}} < \Delta_{T_{\text{approx}}} \text{ OK}$$

Results

Use an 6 inch × 16 inch beam.

Sample Problem 7-6: Determine if the deflection of the steel beam shown in Figure S7-9 is within the deflection limits for total and live load deflection. Use $I = 958 \text{ in}^4$ and $E = 29 \times 10^3$ ksi.

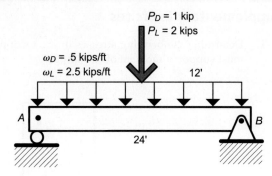

$P_D = 1$ kip
$P_L = 2$ kips

$\omega_D = .5$ kips/ft
$\omega_L = 2.5$ kips/ft

12'

A

B

24'

FIGURE S7-9

SOLUTION

Step 1   Establish the limits of total and live load deflection.

$$\Delta_{T_{\text{Allowable}}} = \frac{\ell}{240} = \frac{24 \text{ ft}(12 \text{ in/ft})}{240} = 1.20 \text{ in}$$

$$\Delta_{L_{\text{Allowable}}} = \frac{\ell}{360} = \frac{24 \text{ ft}(12 \text{ in/ft})}{360} = 0.80 \text{ in}$$

Step 2   Determine total load deflection using the formulas in Table 7-1(a) and 7-1(b).

$$P_T = P_D + P_L = 1 \text{ kip} + 2 \text{ kips} = 3 \text{ kips}$$

$$\omega_T = \omega_D + \omega_L = .5 \text{ kips/ft} + 2.5 \text{ kips/ft} = 3 \text{ kip/ft}$$

$$\Delta_{T_{\text{actual}}} = \left(\frac{P_T L^3}{48 \, EI} + \frac{5 \, \omega_T L^4}{384 \, EI} \right) 1,728$$

$$= \left(P_T + \frac{5 \, \omega L}{8} \right) \left(\frac{L^3 1,728}{48 EI} \right)$$

$$= \left[3 \text{ kips} + \frac{5(3 \text{ kip/ft})(24 \text{ ft})}{8} \right] \left[\frac{(24)^3 \, 1,728}{48(29 \times 10^3 \text{ psi})(958 \text{ in}^4)} \right]$$

$$= .86 \text{ in} < 1.2 \text{ in OK}$$

$$\Delta_{L_{actual}} = \left(P_L + \frac{5\omega L}{8}\right)\left(\frac{L^3 1{,}728}{48EI}\right)$$

$$= \left[2 \text{ kips} + \frac{5(2.5 \text{ kip/ft})(24 \text{ ft})}{8}\right]\left[\frac{(24 \text{ ft})^3 1{,}728}{48(29 \times 10^3 \text{ psi})(958 \text{ in}^4)}\right]$$

$$= .71 \text{ in} > .8 \text{ in OK}$$

Results

The beam deflection is within the limits for both dead load and live load deflection.

Supplementary Exercises

7-1. Determine the bending stresses in the sections shown in Figure E7-1. The beam must support a moment of 8,000 ft-lbs.

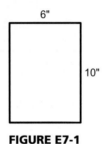

6"

10"

FIGURE E7-1

7-2. What is the maximum concentrated load, (*P*), that the beam shown in Figure E7-2 can support if $F_b = 1{,}400$ psi?

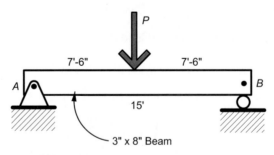

P

7'-6"   7'-6"

A   B

15'

3" x 8" Beam

FIGURE E7-2

7-3. The T-beam shown in Figure E7-3 must support a shear force of 13 kips.
  a. What is the maximum shear stress (f_v) in the section?
  b. What is the shear stress at the top of the web?

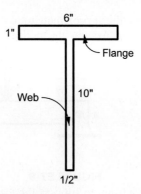

FIGURE E7-3

7-4. A 6 inch × 14 inch beam supports the loads shown in Figure E7-4. Determine the maximum shear stress.

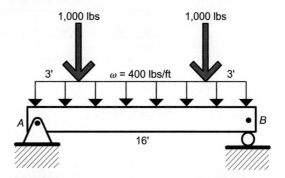

FIGURE E7-4

7-5. Calculate the total load deflection for a 2 inch × 10 inch beam, spanning 16 feet, that supports a uniform distributed load of 120 lbs/ft. Use $E = 1.5 \times 10^6$ psi.

7-6. Determine if the deflection of the steel wide flange beam shown in Figure E7-5 falls within the allowable limits for total and live load deflection. Use $I_x = 1,140$ in^4 and $E = 29 \times 10^3$ ksi.

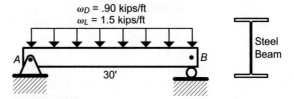

FIGURE E7-5

7-7. Calculate the approximate total deflection for the 8 inch × 12 inch beam shown in Figure E7-6. Assume the modulus of elasticity is 100,000 psi.

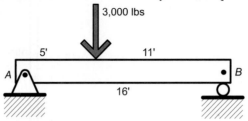

FIGURE E7-6

7-8. Design a beam to support the loads shown in Figure E7-7. The beam must have a rectangular cross section in full inches with the following allowable stresses: F_b = 1,200 psi, F_v = 100 psi and E = 1.4 × 10⁶ psi. Limit the total load deflection to $\ell/240$ and the live load deflection to $\ell/360$.

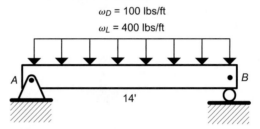

FIGURE E7-7

7-9. Design a beam to support the loads shown in Figure E7-8. The beam must have a rectangular cross section in full inches with the following allowable stresses: F_b = 1,000 psi, F_v = 110 psi, and E = 1.3 × 10⁶ psi. Limit the total deflection to $\ell/240$.

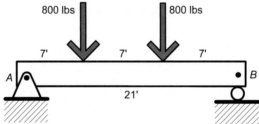

FIGURE E7-8

7-10. Design a beam to support the loads shown in Figure E7-9. The beam must have a rectangular cross section in full inches with the following allowable stresses: F_b = 1,300 psi, F_v = 120 psi, and E = 1.3 × 10⁶ psi. Limit the total deflection to $\ell/240$ and the live load deflection to $\ell/360$.

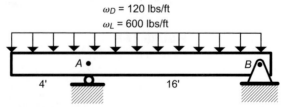

FIGURE E7-9

CHAPTER 8

Columns

8.1 GENERAL

Now that we have learned about beams, we can enhance our knowledge of structures by analyzing columns. As mentioned in Chapter 4, the load that a column, or compression member, can support is controlled by its length. The two basic types of columns, short columns and long columns, are distinguished by their mode of failure, which results from their length and the load they must support. Short columns when subjected to heavy loads will fail by crushing, which is a material failure. Long columns when subjected to heavy loads fail by elastic buckling, which is a sideways deflection along the length of the column similar to that discussed in Chapter 7 (7.4 Lateral Stability). If the load is not released, elastic buckling will eventually result in a material failure due to bending. Once again we will call on our friendly gorillas to demonstrate elastic buckling.

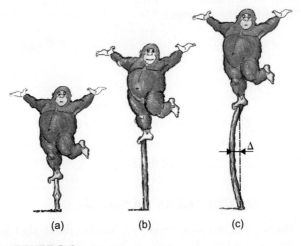

(a) (b) (c)

FIGURE 8-1

The gorilla in Figure 8-1(a) is standing on a short column that apparently does not have sufficient cross-sectional area to support the gorilla's weight and is crushing. The column supporting the gorilla in (b) has sufficient length and cross section to support the load of his weight. If the column becomes longer, as shown in Figure 8-1(c), the same gorilla's weight would cause the column to buckle and eventually collapse.

8.1.1 Short Columns

Stresses in short columns are simple to calculate since they are determined by the basic stress equation, which we learned in Chapter 4.

$$f_c = \frac{P}{A} < F_c$$

where:
f_c = the actual compressive stress,
P = the actual axial load,
A = the area of the column cross section, and
F_c = the allowable compressive stress.

We can safely state that short columns are compression members rather than columns. In short columns or compression members, the failure is due to the yield stress (F_y) of the material, which is less than the *critical buckling stress*, which we discuss later in this chapter.

8.1.2 Long Columns

The behavior of long columns within the elastic limit was first investigated by Leonhard Euler, a Swiss mathematician, in the 18th century. He developed an equation, which in its modified form is still in use today. Euler's equation presents the relationship between the load that causes buckling of a pinned end column and the stiffness of the column. The critical load that causes buckling can be determined by the equation:

$$P_{cr} = \frac{\pi^2 E \, I_{min}}{\ell^2}$$

where:
P_{cr} = the critical buckling load,
E = the modulus of elasticity of the material,
I_{min} = the minumum moment of inertia of the column cross section, and
ℓ = the legth of the column.

8.2   SLENDERNESS RATIO

The term *radius of gyration* (r) is used in the design and analysis of columns. The radius of gyration is a concept that expresses the relationship between the area of a cross section and a centroidal moment of inertia. It measures resistance to buckling about a certain axis and accounts for the moment of inertia, (l), and the cross sectional area (A), which is expressed below.

$$r_x = \sqrt{\frac{I_x}{A}} \quad \text{and} \quad r_y = \sqrt{\frac{I_y}{A}}$$

To develop the slenderness ratio ($\ell/_r$) we will solve for l in the above equation and substitute our value into Euler's equation.

$$r = \sqrt{\frac{I}{A}}$$

$$I = r^2 A$$

$$P_{cr} = \frac{\pi^2 EI}{\ell^2} = \frac{\pi^2 E(Ar^2)}{\ell^2} = \frac{\pi^2 EA}{(\ell/r)^2}$$

Using the formula above we can also develop an equation for the critical stress (f_{cr}).

$$f_{cr} = \frac{P_{cr}}{A} = \frac{\pi^2 EA}{A(\ell/r)^2} = \frac{\pi^2 E}{(\ell/r)^2}$$

It is very important to note that the higher the value of slenderness ratio (ℓ/r), the lower the critical load capacity of the column. As an example, we will calculate the critical load on a 4 inch $\times$ 4 inch column with a modulus of elasticity of 1,500,000 psi and a column length of 10 feet.

$$I = 21.33 \text{ in}^4 \, (\text{Appendix A})$$

$$P_{cr} = \frac{\pi^2 EI_{min}}{\ell^2}$$

$$= \frac{\pi^2 (1.5 \times 10^6 \text{ psi})(21.33 \text{ in}^4)}{(10 \text{ ft} \times 12 \text{ in/ft})^2}$$

$$= 21{,}932.4 \text{ lbs} \approx 21.9 \text{ kips}$$

If we keep the same column but extend its length to 11 feet, we will see how the critical load is decreased from 21.9 kips to 18.1 kips.

$$P_{cr} = \frac{\pi^2 EI_{min}}{\ell^2}$$

$$= \frac{\pi^2 (1.5 \times 10^6 \text{ psi})(21.33 \text{ in}^4)}{(11 \text{ ft} \times 12 \text{ in/ft})^2}$$

$$= 18{,}123.2 \text{ lbs} \approx 18.1 \text{ kips}$$

By extending the column length by one foot we decrease the critical load by 3.8 kip.

8.3 EFFECTIVE LENGTH

When Euler developed his equation, he did so using a column that was pinned at the top and bottom with the load placed at the centroid of the column's cross section. From Chapter 2 we learned that there are connection types other than pinned, such as fixed connections. Each type of connection affects the column's behavior by changing the *effective length* (ℓ_e) of the column, the length used in determining the critical load. If we use the column with a pinned connection as our standard, we find that the column with a pinned connection has ℓ equal to ℓ_e. From that point we develop values for ℓ_e for columns with various types of connections.

Figure 8-2(a) illustrates a column with both ends pinned—with (ℓ_e) equal to ℓ. In Figure 8-2(b) one end of the column is fixed, while the other remains pinned. The effective length in this situation changes to .7 ℓ. It does not make any difference if the fixed end is on the top or the bottom; we would still use the same effective length. Figure 8-2(c) shows a column with both ends fixed. In this instance we would use an effective length of .5 ℓ.

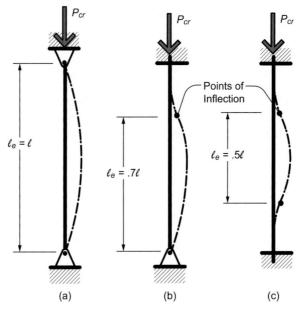

FIGURE 8-2

Note:

The values in Figure 8-2 are theoretical and less conservative than those used in practice. Values used in practice have safety factors and take into consideration defects and other characteristics of the material of which the column is made.

Having established the effects of the end connections, we can adjust Euler's equation to reflect these conditions.

$$P_{cr} = \frac{\pi^2 E I_{min}}{\ell_e^2}$$

To show how ℓ_e changes the critical load capacity of the column, we assume a 3 inch × 4 inch, 12-foot-long column with $E = 1.2 \times 10^6$ psi, first with the ends pinned shown in Figure 8-3.

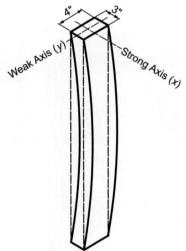

FIGURE 8-3

The column is weaker about the 3 inch dimension, and it will buckle about that side before it will buckle about the 4 inch dimension. This makes the 3 inch side of the column or y axis the weak axis, so we select the moment of inertia and radius of gyration about this y axis.

$$I_y = 9.00 \text{ in}^4 \text{ (from Appendix A)}$$

$$\ell_e = \ell = (12 \text{ ft})(12 \text{ in/ft}) = 144 \text{ in}$$

$$P_{cr} = \frac{\pi^2 EI_{min}}{\ell_e^2} = \frac{\pi^2 (1.2 \times 10^6 \text{ psi})(9.00 \text{ in}^4)}{(144 \text{ in})^2} = 5,140.4 \text{ lbs}$$

We will now place a fixed connection on the column and recalculate the critical load.

$$I_y = 9.00 \text{ in}^4 \text{ (from Appendix A)}$$

$$\ell_e = .7\ell = .7(12 \text{ ft})(12 \text{ in/ft}) = 100.8 \text{ in}$$

$$P_{cr} = \frac{\pi^2 EI_{min}}{\ell_e^2} = \frac{\pi^2 (1.2 \times 10^6 \text{ psi})(9.00 \text{ in}^4)}{(100.8 \text{ in})^2} = 10,490.7 \text{ lbs}$$

By fixing one end of the column, the critical load capacity is increased by 100%.

8.4 LATERAL BRACING

Now we will examine what will happen if a girt (a horizontal member placed between the exterior column's wall cladding) is placed at mid-height of a column on the weak the axis (Figure 8-4 on page 212).

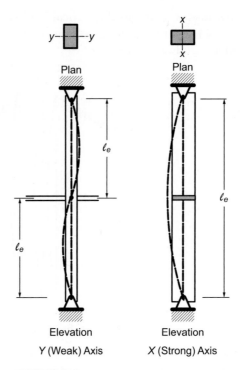

Plan

Plan

Elevation

Elevation

Y (Weak) Axis

X (Strong) Axis

FIGURE 8-4

When bracing is used each axis must be checked to determine which one is the weakest (larges ℓ_e) and will buckle first. To show how to calculate P_{cr} for the column in Figure 8-4, we will assume the column is a 3 inch × 5 inch rectangular cross section with a modulus of elasticity of 1.6×10^6 psi and a length of 14 feet.

Weak Axis (y − y)

$$I_y = 11.25 \text{ in}^4 \text{ (from Appendix A)}$$
$$r_y = .87 \text{ in}$$
$$\ell_{ey} = 7 \text{ ft} \times 12 \text{ in/ft} = 84 \text{ in}$$
$$\text{Slenderness Ratio} = (\ell_{ey}/r_y) = (84 \text{ in}/.87 \text{ in}) = 96.55$$

Strong Axis (x − x)

$$l_x = 31.25 \text{ in}^4$$
$$\ell_{ex} = 14 \times 12 \text{ in/ft} = 168 \text{ in}$$
$$r_x = 1.44 \text{ in}$$

Slenderness Ratio $= (\ell_{ex}/r_x) = (168 \text{ in}/1.44 \text{ in}) = 116.67 \leftarrow$ CONTROLS

$$P_{cr} = \frac{\pi^2 EI}{(\ell_e)^2} = \frac{\pi^2 (1.6 \times 10^6 \text{ psi})(31.25 \text{ in}^4)}{(168 \text{ in})^2} = 17{,}484 \text{ lbs}$$

Even though we would suspect a 3 inch × 5 inch column to buckle about the y axis or the 3 inch side, this problem proves that the column would actually buckle about the strong axis due to the bracing of the weak axis.

Sample Problems

Sample Problem 8-1: Determine the critical buckling stress for a column 89 × 89 mm in cross section and 3 m in length. Assume $E = 12{,}000$ MPa.

SOLUTION

Step 1 Using SI units we begin by calculating the moment of inertia, radius of gyration, and slenderness ratio.

$$I_x = I_y = \frac{bh^3}{12} = \frac{89 \text{ mm}(89 \text{ mm})^3}{12} = 5.23 \times 10^6 \text{ mm}^4$$

$$r = \sqrt{\frac{I}{A}} = \sqrt{\frac{5.23 \times 10^6 \text{ mm}^4}{7{,}921 \text{ mm}^2}} = 25.7 \text{ mm}$$

$$\ell_e = \frac{\ell}{r} = \frac{3{,}000 \text{ mm}}{25.7 \text{ mm}} = 116.8$$

Step 2 Having all the necessary ingredients, we can now calculate the critical buckling stress.

$$f_{cr} = \frac{\pi^2 E}{(\ell/r)^2} = \frac{\pi^2 (12{,}000 \text{ MPa})(10^3)}{(116.8)^2} = 8{,}682 \text{ kPa}$$

Results

The 89 × 89 mm column will buckle when the critical stresses reach 8,682 kPa.

Sample Problem 8-2: Determine the length at which a 4 inch × 6 inch column will buckle if it supports a load of 40 kip. Assume pinned connections at the top and bottom and $E = 1.8 \times 10^6$.

SOLUTION

Step 1 Using Appendix A, select the moment of inertia about the weak axis for a 4 inch × 6 inch section.

$$I_y = 32.00 \text{ in}^4$$

Step 2 Using the value for the moment of inertia, we can calculate the length of the column using the critical load equation. Care must be exercised in the use of units. The modulus of elasticity is given in lbs/in² and the critical load in kips. To ensure our accuracy the E must be converted to kips/in².

$$P_{cr} = \frac{\pi^2 EI}{(\ell)^2}$$

$$\ell = \sqrt{\frac{\pi^2 EI}{P_{cr}}} = \sqrt{\frac{(\pi^2)(1.8 \times 10^3 \text{ ksi})(32 \text{ in}^4)}{40 \text{ kips}}} = 119.22 \text{ in} = 9.9 \text{ ft}$$

Results

The 4 inch × 6 inch will buckle about the 4 inch side or y axis when a load of 40 kips is applied.

Sample Problem 8-3: Determine the critical buckling load of the column shown in Figure S8-1 below. Assume the column is pinned at the top and $E = 2.4 \times 10^6$ psi.

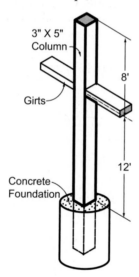

3" X 5"
Column

8'

Girts

12'

Concrete
Foundation

FIGURE S8-1

SOLUTION

Step 1 Determine how the column will buckle by calculating the slenderness ratio for both axes. Care must be exercised in selecting the appropriate ℓ_e, since the column is fixed at its base.

Strong Axis

$$\ell_e = .7(20 \text{ feet})(12 \text{ in/ft}) = 168.0 \text{ in}$$

$$r_x = 1.44 \text{ in}$$

$$\frac{\ell_e}{r_x} = \frac{168.0 \text{ in}}{1.44 \text{ in}} = 116.67 \leftarrow \text{CONTROLS}$$

Weak Axis

$$\ell_e = .7(12 \text{ feet})(12 \text{ in/ft}) = 100.8 \text{ in}$$

$$r_u = .87 \text{ in}$$

$$\frac{\ell_e}{r_y} = \frac{100.8 \text{ in}}{.87 \text{ in}} = 115.86$$

Step 2 Now that we know the strong axis controls, we can calculate the critical buckling load.

$$I_x = 31.25 \text{ in}^4 \text{ (Appendix A)}$$

$$P_{cr} = \frac{\pi^2 EI}{\ell_e^2} = \frac{\pi^2 (2.4 \times 10^6 \text{ psi})(31.25 \text{ in}^4)}{(168)^2} = 26{,}226.6 \text{ lbs} \approx 26.2 \text{ kips}$$

Result

We would normally expect the weak axis to control the buckling, but in this case the lateral bracing shifts the control to the strong axis.

Sample Problem 8-4: A 2 inch diameter column is pinned at the top and bottom. Assume a yield stress (F_y) of 36 ksi and a modulus of elasticity of 29×10^6 ksi, and determine the transition point between the short and long column.

SOLUTION

Step 1 A short column fails when it reaches yield stress and before it reaches critical stress. Realizing this, we can calculate the load on the column using the yield stress.

$$P_{cr} = AF_y$$

$$A = \frac{\pi d^2}{4} = \frac{\pi (2 \text{ in})^2}{4} = 3.14 \text{ in}^2$$

$$P_{cr} = 3.14 \text{ in}^2 (36 \text{ ksi}) = 113.10 \text{ kips}$$

Step 2 Using the load calculated above, we can determine the length of column where the buckling occurs.

$$I = \frac{\pi d^4}{64} = \frac{\pi (2 \text{ in})^4}{64} = .79 \text{ in}^4$$

$$P_{cr} = \frac{\pi^2 EI}{(\ell_e)^2}$$

$$\ell_e = \sqrt{\frac{\pi^2 EI}{P_{cr}}}$$

$$= \sqrt{\frac{\pi^2(29 \times 10^3 \text{ ksi})(.79 \text{ in}^4)}{(113.1 \text{ kips})^2}} = 44.6 \text{ in} = 3.7 \text{ ft}$$

Result

With lengths greater than 3.7 feet, the column will become a long column.

Supplementary Exercises

8-1. Determine the critical buckling stress (f_{cr}) of a 4 inch $\times$ 4 inch column. The column is 12 feet high, pinned at the top and bottom, and the material has a modulus of elasticity of 1.4×10^6 psi.

8-2. Determine the critical buckling load (P_{cr}) for the pipe column shown in Figure E8-1. Use $E = 29 \times 10^3$ ksi, a length of 20 feet, and assume the column is pinned at the top and bottom.

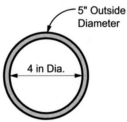

5" Outside Diameter

4 in Dia.

FIGURE E8-1

8-3. A the column shown in Figure E8-2 has its base fixed in a concrete foundation. The top is pinned. Determine the critical buckling load using a height of 18 feet, and assuming the material has a modulus of elasticity of 1.2×10^6.

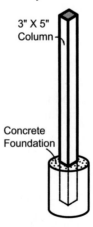

3" X 5" Column

Concrete Foundation

FIGURE E8-2

8-4. The column shown in Figure E8-3 is two stories high and is braced on its weak axis at mid-height. Assume $E = 1.3 \times 106$ psi and determine the critical load the column can support.

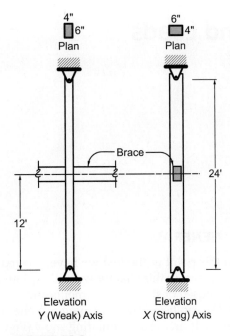

FIGURE E8-3

8-5. Determine the critical stress in the tube column shown in Figure E8-4. The column has a height of 28 feet; it has a fixed connection at the top and a pinned connection at the bottom. The material has a modulus of elasticity of 29×10^3 ksi.

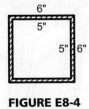

FIGURE E8-4

8-6. Determine the transition point between a short and long column for a 3 inch $\times$ 3 inch column, pinned at the top and bottom. Assume a yield stress (F_y) of 1,210 psi and a modulus of elasticity of $.9 \times 10^6$ psi.

CHAPTER 9

Systems and Loads

9.1 GENERAL

Up to this point in the text we have covered structural elements, trusses, beams, and columns. In this chapter we will explore a few basic structural systems, which are composed of structural elements and the path the loads take to get back to the earth. We can begin by examining a shelter used by family of gorillas who use it for refuge from the accumulating snow (Figure 9-1).

FIGURE 9-1

First, we will remove the roof deck so we can become familiar with common terminology (Figure 9-2) and do some load tracing.

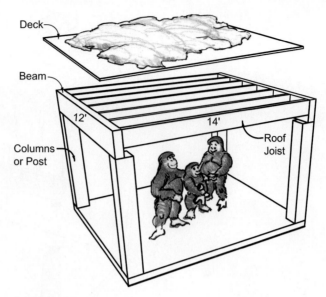

FIGURE 9-2

The family of gorillas is protected from the accumulating snow by the roof deck. The deck is very thin and has a low moment of inertia, so it can't span 12 feet. As the deck deflects, it distributes the snow load along with its own weight to the roof joist. The roof joists, having greater depth (and higher moment of inertia), are capable of spanning 14 feet and supporting the load. The roof joists transfer their load to the beams, which have even greater depth and are capable of supporting half of the total load on the shelter's roof. The last step on the load path journey is taken by the columns, which support half the beam loads and transfer them to the foundation and back to the earth. We can get a much clearer understanding of these load paths by rotating the shelter 90 degrees and illustrating these load paths (Figure 9-3 on page 220). Having a clear picture of how the loads travel through the structure, we can begin an analysis of the system and establish a design methodology for the structural elements.

In areas where there is a possibility of snow accumulation we have to consult the building codes to determine the appropriate snow load. For this example we will assume that the total load on the shelter roof is a snow load (S) of 30 psf and the dead load (D) of 12 psf, which is the weight of the structure.

In Chapter 2, we learned that uniform distributed loads are the result of tributary width times the total load (T). Assuming the roof joists are spaced at 2 feet on center, we can calculate the distributed load in the following manner.

$$T = S + D = \frac{30 \text{ lbs}}{\text{ft}^2} = \frac{12 \text{ lbs}}{\text{ft}^2} = \frac{42 \text{ lbs}}{\text{ft}^2}$$

$$\omega = T \times \text{Tributary width} = \left(\frac{24 \text{ in}}{12 \text{ in/ft}}\right)\left(\frac{42 \text{ lbs}}{\text{ft}^2}\right) = \frac{84 \text{ lbs}}{\text{ft}}$$

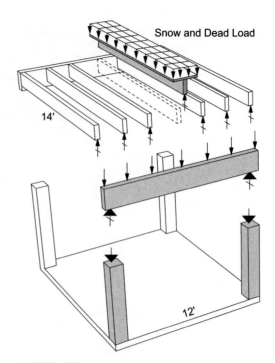

FIGURE 9-3

From the equations provided in Table 7-1(a), we can calculate the reactions.

$$R = \frac{\omega L}{2} = \left(\frac{84 \text{ lbs}}{\text{ft}}\right)\left(\frac{14 \text{ ft}}{2}\right) = 588.0 \text{ lbs}$$

Having calculated the loads and reactions, we can draw the load diagram shown in Figure 9-4.

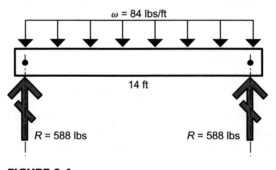

FIGURE 9-4

Using the equations in Table 7-1(a), we are now able to design the roof joist. The reactions for the exterior roof joist would be approximately half of the value shown above since they support half of the tributary width of the interior joist.

Working from the top of the roof down, we note that the beam supports the roof joist; therefore, the reactions from the joist become the loads on the beam. The beam supports five loads, symmetrically placed, plus two exterior joist reactions. The beam reactions are calculated as half the loads.

$$R = \frac{6P}{2} = \frac{6(588 \text{ lbs})}{2} = 1,764.0 \text{ lbs}$$

The load diagram is shown in Figure 9-5.

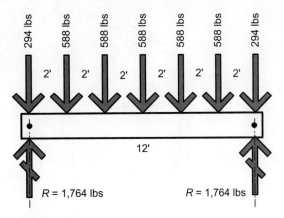

FIGURE 9-5

Clearly, this loading diagram as illustrated is mathematically cumbersome. We would, using Table 7-1, have to calculate the moment, shear, and deflection for a concentrated load in the center and two pair of forces of equal magnitude and symmetrically placed. We would ignore the two exterior forces since they do not influence the moment, shear, and deflection.

Let us see if we can find a simpler way of designing the beam. If we note that the reactions are half of the roof load, we can use this information to create a uniform distributed load.

$$\omega = T \times \text{tributary width} = \left(\frac{42 \text{ lbs}}{\text{ft}^2} \right)\left(\frac{14 \text{ ft}}{2} \right) = 294 \text{ lbs/ft}$$

Using the uniform distributed load, the reactions are half of W.

$$W = \omega L_L = \left(\frac{294 \text{ lbs}}{\text{ft}} \right) 12 \text{ ft} = 3,528.0 \text{ lbs}$$

$$R = \frac{W}{2} = \frac{3,528.0 \text{ lbs}}{2} = 1,764 \text{ lbs}$$

Using the uniform distributed load, we get the same reactions as using the concentrated loads. By using the equation given in Table 7-1(a), we get a moment of 5,292 ft-lbs. This is the same moment that we would calculate if we used the cumbersome method of concentrated loads just described. This simplified method of using distributed loads can be used on any joist, beam, or girder supporting three or more concentrated loads.

If we change the construction of the shelter roof to sloped members, both the terminology and load calculation change. Figure 9-6 provides the terminology.

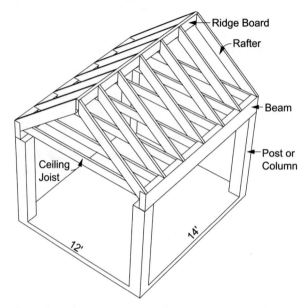

FIGURE 9-6

We will continue to examine the dead and snow loads on these members by extracting a pair of rafters and a ceiling joist (Figure 9-7 on page 223). On any sloped member the dead load is calculated along its actual length, while the snow or roof live loads are calculated along the horizontal projection of the member. With any sloped member the dead load should be converted to horizontal projection before being combined with snow or roof live load. Using the same loads as the previous example, we can calculate this conversion. Detemine the c side of the slope.

$$c = \sqrt{a^2 + b^2} = \sqrt{10^2 + 12^2} = 15.62$$

$$\ell_{horizontal} = 6 \text{ ft}$$

$$\frac{\ell_{actual}}{6 \text{ ft}} = \frac{15.62}{12}$$

$$\ell_{actual} = 7.81 \text{ ft}$$

$$\omega D_{actural} = D \times \text{tributary width} = \frac{12 \text{ lbs}}{ft^2} \times 2 \text{ ft} = 24 \text{ lbs/ft}$$

$$\omega D_{horizontal} = \omega D_{actural} \times \frac{\ell_{actual}}{\ell_{horizontal}} = 24 \text{ lbs/ft} \left(\frac{7.81 \text{ ft}}{6 \text{ ft}} \right) = 31.24 \text{ lbs}$$

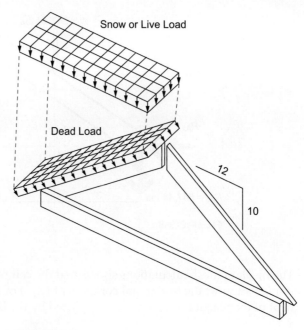

Snow or Live Load

Dead Load

12

10

FIGURE 9-7

Having the dead load on the horizontal projection, we can combine it with the snow load, and calculate the rafter reactions.

$$\omega_s = S \times \text{tributary width} = \left(\frac{30 \text{ lbs}}{\text{ft}^2} \right) 2 \text{ ft} = 60 \text{ lbs/ft}$$

$$\omega_T = \omega_s + \omega_D = 60 \text{ lbs/ft} + 31.24 \text{ lbs/ft} = 91.24 \text{ lbs/ft}$$

When we examine Figure 9-6, we realize the ridge board cannot support loads since it has no vertical support; therefore, we have to exercise extreme care in calculating the reactions of the rafters. We assume the total vertical load is supported by the beams at the lower end of the rafter. The horizontal reaction at the top of the rafter is transferred to the rafter on the other side of the roof.

$$R_{Ay} = \omega_T \times \ell_{\text{horizontal}} = 91.24 \text{ lbs/ft}(6 \text{ ft}) = 547.44 \text{ lbs}$$

Using Figure 9-8 on page 224, the remaining reactions can now be calculated. Since $W = R_{Ay}$, we have only to calculate R_B and R_{Ax}.

$$\Sigma M_A = 0$$

$$= -W(3 \text{ ft}) + R_B(5 \text{ ft})$$

$$R_B = \frac{547.44 \text{ lbs}(3 \text{ ft})}{5 \text{ ft}} = 328.46 \text{ lbs}$$

$$\Sigma F_x = 0$$

$$= -328.46 \text{ lbs} + R_{Ay}$$

$$R_{Ay} = 328.46 \text{ lbs}$$

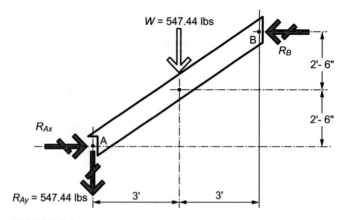

FIGURE 9-8

The result of these calculations shows that the ceiling joist must be in tension since it must support the horizontal component (R_{Ax}) of the rafter. The vertical component of the reaction R_{Ay} must be supported by the beam (Figure 9-9).

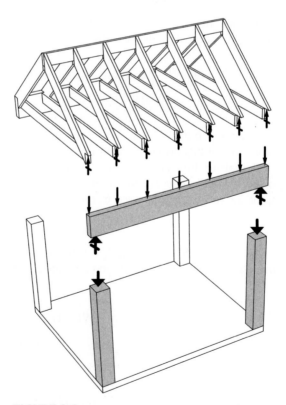

FIGURE 9-9

Using the method described above, we can calculate the loads on the beam using a uniform distributed load.

$$T = \frac{91.24 \text{ lbs/ft}}{2 \text{ ft}} = \frac{45.62 \text{ lbs}}{\text{ft}^2}$$

$$\omega_T = T \times \text{tributary width} = \left(\frac{45.62 \text{ lbs}}{\text{ft}^2}\right) 6 \text{ ft} = 273.72 \text{ lb/ft}$$

Having the uniform distributed load, we can calculate the loads on the post by calculating the beam reactions.

$$R = \frac{\omega_T L}{2} \text{ (Table 7-1a)}$$

$$= \frac{273.72 \text{ lbs/ft}(14 \text{ ft})}{2} = 1{,}916 \text{ lbs}$$

Now that we have R, which is the load on the column, we could assume column size and see if it is below Euler's critical buckling load (see Sample Problem 9-1).

9.2   LATERAL LOADS

In Chapter 1 (1.7, Introduction to Loads), we referred to lateral loads. The calculations of these loads are beyond the scope of this text, but it is essential that we have some understanding of the effects that wind and earthquake (seismic) loads have on structures. To help with this, we will introduce the last of our gorillas whose name is *Lat*, short for Lateral Force. Lat loves to push on buildings until they deflect excessively or overturn. The only way we can prevent this from happening is to install a *lateral force resisting system* (LRFS). We will let Lat loose on a building without a LRFS (Figure 9-10).

FIGURE 9-10

Lat is doing a great job. The roof is separating from the columns and structure is about to collapse.

To compensate for this, we will now add shear wall to the structure. These walls must be constructed of concrete, masonry, steel, wood, or any material that is rigid and capable of accepting a great deal of shear (Figure 9-11).

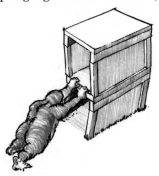

FIGURE 9-11

Lat can push as much as he wants but he can not overturn or collapse the structure. It will deform, but we can control the lateral deflection (in this case, called drift) by the design of the shear walls.

Another option is to replace the shear walls with bracing. In Figure 9-12 we used *X-bracing*. There are other forms of bracing that take the shape of a K or a chevron. The selection of the bracing system depends on the loads and the architecture we are trying to accommodate. We want to produce a stable structure while not interfering with the location of doors and windows or other architectural elements.

FIGURE 9-12

Our last LFRS is a rigid frame. Should shear wall or braced frames interfere with the architectural intent, rigid frames provide an alternate solution (Figure 9-13).

FIGURE 9-13

In Chapter 2 we learned about *fixed connections*. Rigid frames are constructed of rigid connections. All the connections of beams to columns are rigidly connected.

In summary, there are three basic systems: shear walls, braced frames, and rigid frames. The order in which these are listed is the order of stiffness with shear walls providing the greatest stiffness and rigid frames the least stiffness.

Most buildings employ more than one of these systems due to the form and configuration of the plans. **The most important issue presented here is that all buildings are subjected to wind and/or earthquake forces and, therefore, all building structures must have a lateral force resisting system**. Remember Lat is always present whether we are designing, fabricating, or constructing structural systems.

We will now say goodbye to Lat and the rest of our friendly gorillas. It is time to move on to more advanced study of structural materials such as wood, steel, concrete, and masonry.

Sample Problems

Sample Problem 9-1: We will conclude Chapter 9 with a problem about loads and load paths, this time carrying each step through the design process. It is important that we realize this is an academic exercise and is performed in order to prepare us for advanced studies in structural materials. We will begin by reviewing the roof framing shown in Figure S9-1.

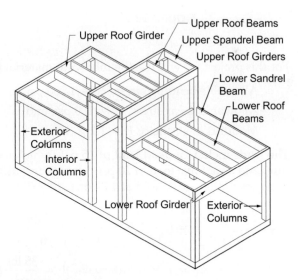

FIGURE S9-1

The dimensions of the system are shown in Figure S9-2 and Figure S9-3.

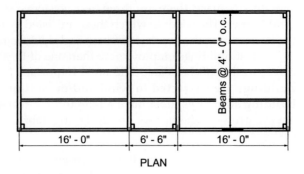

PLAN

FIGURE S9-2

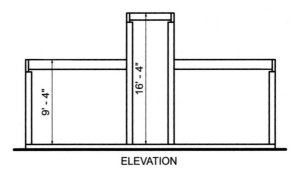

ELEVATION

FIGURE S9-3

For this project we will use the following material properties:

Allowable bending stress, F_b = 1,200 psi,

Allowable shear stress, F_v = 100 psi, and

Modulus of elasticity, E = 1.2 × 10^6 psi.

The loads include a dead load of 15 psf (which includes the weight of the structure) and a roof live load of 20 psf. We can start at the top of the structure with the *upper beams* that are spaced at 4-0″ on center (Figure S9-1).

$$T = D + L_r = \frac{15 \text{ lbs}}{\text{ft}^2} + \frac{20 \text{ lbs}}{\text{ft}^2} = 35 \text{ psf}$$

$$\omega_T = T \times \text{tributary width} = \frac{35 \text{ lbs}}{\text{ft}^2}(4 \text{ ft}) = 140 \text{ lb/ft}$$

Using the formulas in Table 7-1a, make a trial selection based on moment.

$$M = \frac{\omega L2}{8} = \frac{140 \text{ lbs/ft}(6.5 \text{ ft})^2}{8} = 739 \text{ ft-lbs}$$

$$S_{\text{req'd}} = \frac{M}{F_b} = \frac{739 \text{ ft-lbs}(12 \text{ in/ft})}{1,200 \text{ lbs/in}^2} = 7.39 \text{ in}^3$$

Try: 2×5 (Appendix A) $S_x = 8.33 \text{ in}^3$

After we make our trial selection from the Properties Table (Appendix A), we should list the other important properties we will need to complete our calculations. We will need the cross-sectional area, $A = 10 \text{ in}^2$, to check the shear and the moment of inertia, $I_x = 20.83 \text{ in}^4$, to check the deflection.

Having calculated the trial selection, we now check the shear.

$$V = \frac{wL}{2} = \frac{(140 \text{ lbs/ft})(6.5 \text{ ft})}{2} = 455 \text{ lbs}$$

$$F_v = 100 \text{ psi}$$

$$f_v = \frac{3V}{2A} = \frac{3(455 \text{ lbs})}{2(10 \text{ in}^2)} = 68.25 \text{ psi} < 100 \text{ psi} \quad \text{OK}$$

Check deflection.

$$\Delta_{T \text{ Allowable}} = \frac{\ell}{180} = \frac{6.5 \text{ ft}(12 \text{ in/ft})}{180} = .43 \text{ in}$$

$$\Delta_{L_r \text{ Allowable}} = \frac{\ell}{240} = \frac{6.5 \text{ ft}(12 \text{ in/ft})}{240} = .33 \text{ in}$$

$$\Delta_T = \frac{5 \omega_T L^4 (1,728)}{384 \text{ El}} = \frac{5(140 \text{ lbs/ft})(6.5 \text{ ft})^4(1,728)}{384(1.2 \times 10^6 \text{ psi})(20.83 \text{ in}^4)} = .22 \text{ in} < .43 \text{ in} \quad \text{OK}$$

$$\Delta_T < \Delta_{L_r \text{ Allowable}} \quad (\text{No check for live load deflection required})$$

The 2×5 meets all the requirements for moment, shear, and deflection, so we can state: Use 2×5 @ 4' -0" on center.

The upper girder supports three concentrated loads so we can use the uniform distributed load for design.

$$\omega_T = T \times \text{tributary width} = \frac{35 \text{ lbs/ft}}{\text{ft}^2}(3.25 \text{ ft}) = 113.75 \text{ lbs/ft}$$

$$M = \frac{\omega L^2}{8} = \frac{113.75 \text{ lbs/ft}(16 \text{ ft})^2}{8} = 3,640 \text{ ft-lbs}$$

$$S_{\text{req'd}} = \frac{M}{F_b} = \frac{3{,}640 \text{ ft-lbs}(12 \text{ in/ft})}{1{,}200 \text{ lbs/in}^2} = 36.40 \text{ in}^3$$

Try: $3 \times 9 \; S_x = 40.50 \text{ in}^3 > 36.40 \text{ in}^3 (A = 27 \text{ in}^2, I_x = 182.25 \text{ in}^4)$

Check shear.

$$V = \frac{\omega L}{2} = \frac{(113.75 \text{ lbs/ft})16 \text{ ft}}{2} = 910 \text{ lbs}$$

$$f_v = \frac{3V}{2A} = \frac{3(910 \text{ lbs})}{2(27 \text{ in}^2)} = 50.56 \text{ psi} < 100 \text{ psi} \quad \text{OK}$$

Check deflection.

$$\Delta_T = \frac{\ell}{180} = \frac{16 \text{ ft}(12 \text{ in/ft})}{180} = 1.07 \text{ in}$$

$$\Delta_{L_r} = \frac{\ell}{240} = \frac{16 \text{ ft}(12 \text{ in/ft})}{240} = .80 \text{ in}$$

$$\Delta_T = \frac{5 \, \omega_T L^4 (1{,}728)}{384 \; El}$$

$$= \frac{5(113.75 \text{ lbs/ft})(16 \text{ ft})^4(1{,}728)}{384(1.2 \times 10^6 \text{ psi})(182.25 \text{ in})^4} = .77 \text{ in} < 1.07 \text{ in} \quad \text{OK}$$

$\Delta_T < \Delta_{L_r \text{ Allowable}}$ (No check for live load deflection required)

Use 3×10 upper girder.

Remember that the shear, V, is the reaction and also the force on the top of the column.

The lower portion of the roof system supports the same loads as the upper portion. We begin by designing the beams.

$$T = 35 \text{ psf (previous calculations)}$$

$$\omega_T = 140 \text{ lbs/ft (same spacing as upper beams)}$$

$$M = \frac{\omega L^2}{8} = \frac{140 \text{ lbs/ft}}{8}(16 \text{ ft})^2 = 4{,}480.00 \text{ ft-lbs}$$

$$S_{\text{req'd}} = \frac{M}{F_b} = \frac{4{,}480.00 \text{ ft-lbs}(12 \text{ in/ft})}{1{,}200 \text{ lbs/in}^2} = 44.80 \text{ in}^3$$

Try: $3 \times 10 \; S_x = 50.00 \text{ in}^3 > 44.80 \text{ in}^3 (A = 30 \text{ in}^2, I_x = 250 \text{ in}^4)$

Check shear.

$$V = \frac{\omega L}{2} = \frac{140 \text{ lbs/ft}(16 \text{ ft})}{2} = 1{,}120.00 \text{ lbs}$$

$$f_v = \frac{3V}{2A} = \frac{3(1{,}120.00 \text{ lbs})}{2(30 \text{ in}^2)} = 56 \text{ psi} < 100 \text{ psi} \quad OK$$

Check deflection.

$$\Delta_T = \frac{\ell}{180} = \frac{16 \text{ ft}(12 \text{ in/ft})}{180} = 1.07 \text{ in}$$

$$\Delta_{L_r} = \frac{\ell}{240} = \frac{16 \text{ ft}(12 \text{ in/ft})}{240} = .80 \text{ in}$$

$$\Delta_T = \frac{5\omega_T L^4(1{,}728)}{384 \, EI} = \frac{5(140.00 \text{ lbs/ft})(16 \text{ ft})^4(1{,}728)}{384(1.2 \times 10^6 \text{ psi})(250.00 \text{ in}^4)}$$

$\Delta_T < \Delta_{L_r \, Allowable}$ (No check for live load deffection required)

Use 3 × 10 @ 4'-0" on center.

The only remaining horizontal member we have left to design is the lower girders.

$$T = 35 \text{ psf}$$

$$\omega_T = T \times \text{tributary width} = \frac{35 \text{ lbs}}{\text{ft}^2}(8 \text{ ft}) = 280 \text{ lbs/ft}$$

$$M = \frac{\omega L^2}{8} = \frac{280 \text{ lbs/ft}(16 \text{ ft})^2}{8} = 8{,}960.00 \text{ ft-lbs}$$

$$S_{\text{req'd}} = \frac{M}{F_b} = \frac{8{,}960.00 \text{ ft-lbs}(12 \text{ in/ft})}{1{,}200 \text{ lbs/in}^2} = 89.60 \text{ in}^3$$

Try: 4 × 12 S_x = 96.00 in^3 > 89.60 in^3(A = 48 in^2, I_x = 576 in^4)

Check shear.

$$V = \frac{\omega L}{2} = \frac{280 \text{ lbs/ft}(16 \text{ ft})}{2} = 2{,}240 \text{ lbs}$$

$$fv = \frac{3V}{2A} = \frac{3(2{,}240 \text{ lbs})}{2(48 \text{ in}^2)} = 70 \text{ psi} < 100 \text{ psi} \quad OK$$

Check deflection.

$$\Delta_T = \frac{\ell}{180} = \frac{16 \text{ ft}(12 \text{ in/ft})}{180} = 1.07 \text{ in}$$

$$\Delta_{L_r} = \frac{\ell}{240} = \frac{16 \text{ ft}(12 \text{ in/ft})}{240} = .80 \text{ in}$$

$$\Delta_T = \frac{5\omega_T L^4(1,728)}{384 \text{ El}}$$

$$= \frac{5(280.00 \text{ lbs/ft})(16 \text{ ft})^4(1,728)}{384(1.2 \times 10^6 \text{ psi})(576 \text{ in})^4} = .60 \text{ in} < 1.07 \text{ in}   \text{OK}$$

$$\Delta_T < \Delta_{L_r \text{ Allowable}}$$

(No check for live load defleciton required)

Use 4 × 12 lower girder

The shear from the lower girder is the reaction and the force on the exterior columns, which is 2,240 lbs. The depth of the lower girder is 12 inches so the height of the exterior column is 8'- 4" or 8.33'. We can assume the exterior columns are 4 inch × 4 inch, pinned at the top and bottom, and see if the load is less than the critical buckling load.

$$P = 2,240 \text{ lbs}$$

$$l_x = l_y = 21.33 \text{ in}^4$$

$$P_{cr} = \frac{\pi^2 El_{min}}{\ell^2}$$

$$= \frac{\pi^2(1.2 \times 10^6 \text{ lbs/in}^2)21.33 \text{ in}^4}{(8.33 \text{ ft} \times 12 \text{ in/ft}^2)} = 25,262 \text{ lbs}$$

$$P_{cr} > P   \text{OK}$$

The actual column load is well below the critical buckling load, but we do not want to use a smaller column since the girder it supports is 4 inches wide.

The interior columns present a different problem. These columns support a load of 910 pounds from the upper level girders and 2,240 pounds from the lower level girders, so the total load on the column is 3,150 pounds. Figure S9-1 illustrates that the interior columns are braced on both axes at the same height as the exterior columns; therefore, we can conclude the interior columns are also within the limits of the critical buckling load and the 4 inch × 4 inch interior columns are sufficient.

There is only one thing left to say.

FIGURE S9-4

Supplementary Exercises

Problems 9-1 through 9-5 refer to the partial framing plan shown below. For these problems we will assume the following allowable stresses:

$$F_b = 1200 \text{ psi}$$

$$F_v = 100 \text{ psi}$$

$$E = 1.2 \times 10^6 \text{ psi}.$$

We will also assume the following loads:

Roof: D = 12 psf (includes the weight of the structure)
         S = 40 psf
Floor: D = 15 psf (includes the weight of the structure)
         L = 100 psf.

9-1. Using rectangular cross sections design the interior roof joist limiting the total deflection to L/240 and the snow load deflection to L/360. Limits of Euler's critical buckling load (P_{cr}).

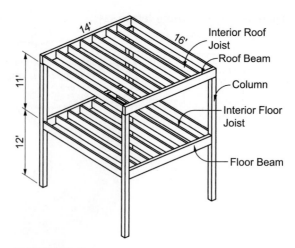

FIGURE E9-1

9-2. Using rectangular cross sections design the roof beam limiting the total defection to L/240 and the snow load deflection to L/360.

9-3. Using rectangular cross sections design the interior floor joist limiting the total defection to L/240 and the live load deflection to L/360.

9-4. Using rectangular cross sections design the floor beam limiting the total deflection to L/240 and the live load deflection to L/360.

9-5. Assume the column is 4" × 4" and determine if the loads on the column are within the limits of Euler's critical buckling load (P_{cr}).

APPENDIX A

Properties of Members with Rectangular Cross Sections

b (in.)	h (in.)	A (in²)	Properties x axis			Properties y axis		
			I_x (in⁴)	S_x (in³)	r_x (in.)	I_y (in⁴)	S_y (in³)	r_y (in)
2	2	4	1.33	1.33	0.58	1.33	1.33	0.58
2	3	6	4.50	3.00	0.87	2.00	2.00	0.58
2	4	8	10.67	5.33	1.15	2.67	2.67	0.58
2	5	10	20.83	8.33	1.44	3.33	3.33	0.58
2	6	12	36.00	12.00	1.73	4.00	4.00	0.58
2	7	14	57.17	16.33	2.02	4.67	4.67	0.58
2	8	16	85.33	21.33	2.31	5.33	5.33	0.58
2	9	18	121.50	27.00	2.60	6.00	6.00	0.58
2	10	20	166.67	33.33	2.89	6.67	6.67	0.58
2	11	22	221.83	40.33	3.18	7.33	7.33	0.58
2	12	24	288.00	48.00	3.46	8.00	8.00	0.58
3	3	9	6.75	4.50	0.87	6.75	4.50	0.87
3	4	12	16.00	8.00	1.15	9.00	6.00	0.87
3	5	15	31.25	12.50	1.44	11.25	7.50	0.87
3	6	18	54.00	18.00	1.73	13.50	9.00	0.87
3	7	21	85.75	24.50	2.02	15.75	10.50	0.87
3	8	24	128.00	32.00	2.31	18.00	12.00	0.87
3	9	27	182.25	40.50	2.60	20.25	13.50	0.87
3	10	30	250.00	50.00	2.89	22.50	15.00	0.87
3	11	33	332.75	60.50	3.18	24.75	16.50	0.87
3	12	36	432.00	72.00	3.46	27.00	18.00	0.87
4	4	16	21.33	10.67	1.15	21.33	10.67	1.15
4	5	20	41.67	16.67	1.44	26.67	13.33	1.15
4	6	24	72.00	24.00	1.73	32.00	16.00	1.15
4	7	28	114.33	32.67	2.02	37.33	18.67	1.15
4	8	32	170.67	42.67	2.31	42.67	21.33	1.15
4	9	36	243.00	54.00	2.60	48.00	24.00	1.15
4	10	40	333.33	66.67	2.89	53.33	26.67	1.15
4	11	44	443.67	80.67	3.18	58.67	29.33	1.15
4	12	48	576.00	96.00	3.46	64.00	32.00	1.15

b (in.)	h (in.)	A (in²)	Properties x axis			Properties y axis		
			I_x (in⁴)	S_x (in³)	r_x (in.)	I_y (in⁴)	S_y (in³)	r_y (in)
5	5	25	52.08	20.83	1.44	52.08	20.83	1.44
5	6	30	90.00	30.00	1.73	62.50	25.00	1.44
5	7	35	142.92	40.83	2.02	72.92	29.17	1.44
5	8	40	213.33	53.33	2.31	83.33	33.33	1.44
5	9	45	303.75	67.50	2.60	93.75	37.50	1.44
5	10	50	416.67	83.33	2.89	104.17	41.67	1.44
5	11	55	554.58	100.83	3.18	114.58	45.83	1.44
5	12	60	720.00	120.00	3.46	125.00	50.00	1.44
5	13	65	915.42	140.83	3.75	135.42	54.17	1.44
5	14	70	1143.33	163.33	4.04	145.83	58.33	1.44
6	6	36	108.00	36.00	1.73	108.00	36.00	1.73
6	7	42	171.50	49.00	2.02	126.00	42.00	1.73
6	8	48	256.00	64.00	2.31	144.00	48.00	1.73
6	9	54	364.50	81.00	2.60	162.00	54.00	1.73
6	10	60	500.00	100.00	2.89	180.00	60.00	1.73
6	11	66	665.50	121.00	3.18	198.00	66.00	1.73
6	12	72	864.00	144.00	3.46	216.00	72.00	1.73
6	13	78	1098.50	169.00	3.75	234.00	78.00	1.73
6	14	84	1372.00	196.00	4.04	252.00	84.00	1.73
6	15	90	1687.50	225.00	4.33	270.00	90.00	1.73
6	16	96	2048.00	256.00	4.62	288.00	96.00	1.73
8	8	64	341.33	85.33	2.31	341.33	85.33	2.31
8	9	72	486.00	108.00	2.60	384.00	96.00	2.31
8	10	80	666.67	133.33	2.89	426.67	106.67	2.31
8	11	88	887.33	161.33	3.18	469.33	117.33	2.31
8	12	96	1152.00	192.00	3.46	512.00	128.00	2.31
8	13	104	1464.67	225.33	3.75	554.67	138.67	2.31
8	14	112	1829.33	261.33	4.04	597.33	149.33	2.31
8	15	120	2250.00	300.00	4.33	640.00	160.00	2.31
8	16	128	2730.67	341.33	4.62	682.67	170.67	2.31

Answers to Supplemental Exercises

CHAPTER 1

1-1. a. $F_x = 86.6$ lbs, $F_y = 50$ lbs
b. $F_x = 2.07$ kips, $F_y = 1.21$ kips

1-2. $R = 340$ lbs at $3.4°$ from the horizon to the Southeast

1-3. $R = 69$ kN at $42°$ from the horizon to the Southwest

1-4. $R = 289$ lbs at $6°$ from the horizon to the Southwest

1-5. $R = 81$ kN at $23°$ from the horizon to the Southeast

1-6. $R = 340.6$ lbs at $3.4°$ from the horizon to the Southeast

1-7. $R = 288.82$ lbs at $6.2°$ from the horizon to the Southwest

1-8. $R = 81.42$ kN at $22.91°$ from the horizon to the Southeast

1-9. $R = 6$ kN at 3.17 m from A

1-10. $M_A = -4.480°$ ft-lbs

1-11. $M_B = 28$ kip-ft

CHAPTER 2

2-1. $R_{Ax} = 0$, $R_{Ay} = 350$ lbs, $R_B = 550$ lbs

2-2. $R_{Ax} = 30$ kN, $R_{Ay} = 65$ kN, $R_B = 35$ kN

2-3. $R_A = 13.33$ kips, $R_{By} = 6.67$ kips, $R_{Bx} = 0$

2-4. $R_A = 2,200$ lbs, $R_{By} = 2,200$ lbs, $R_{Bx} = 0$

2-5. $R_{Ax} = 0$, $R_{Ay} = 24$ kips, $R_B = 36$ kips

2-6. $R_{Ax} = 0$, $R_{Ay} = 35.14$ kN, $R_B = 24.86$ kN

2-7. $R_{Ax} = 0$, $R_{Ay} = 21.5$ kips, $R_B = 18.5$ kips

2-8. $R_{Ax} = 0$, $R_{Ay} = R_{By} = 117$ kips

2-9. $R_A = 54$ kips, $R_{By} = 108$ kips, $R_{Bx} = 0$

2-10. $R_A = 5,700$ lbs, $R_{By} = 7,800$ lbs, $R_{Bx} = 0$

CHAPTER 3

3-1. $R_{Ax} = -300$ lbs, $R_{Ay} = 187.5$ lbs, $R_B = 312.5$ lbs,
$AB = 487.5$ lbs C, $BC = 812.5$ lbs C, $CD = 750$ lbs T, $DA = 750$ lbs T, $DB = 500$ lbs T

3-2. $R_A = 450$ lbs, $R_{By} = 450$ lbs, $R_{Bx} = 0$
$AB = 811.25$ lbs C, $BC = 900$ lbs C, $CD = 900$ lbs C, $DE = 811.25$ lbs C,
$EF = 675$ lbs T, $FD = 0$, $FG = 675$ lbs T, $GD = 270.42$ lbs T, $GC = 300$ lbs C,
$GB = 270.42$ lbs T, $GH = 675$ lbs T, $HB = 0$, $HA = 675$ lbs T

3-3. $R_A = 50$ kN, $R_{By} = 30$ kN, $R_{Bx} = 30$ kN
$AB = 0$, $BC = 30$ kN T, $CD = 7.5$ kN C, $DE = 22.5$ kN C, $EF = 0$, $FG = 0$,
$GH = 52.5$ kN T, $HI = 60$ kN T, $IJ = 60$ kN T, $JA = 37.5$ kN T, $AC = 62.5$ kN C,
$CJ = 30$ kN T, $JD = 37.5$ kN C, $DI = 0$, $DH = 12.5$ kN C, $HE = 30$ kN T,
$EG = 37.5$ kN C

3-4. $R_{Ay} = 4$ kips, $R_{Ax} = 0$, $R_B = 4$ kips
$AB = 1.0$ kip C, $BC = 0$, $CD = -3.16$ kips C, $DE = 6.32$ kips C, $EF = 9.49$ kips C,
$FG = 9.0$ kips T, $GH = 9.0$ kips T, $HI = 6.0$ kips T, $IA = 3.0$ kips T, $AC = 4.24$ kips C,
$CI = 2.0$ kips T, $ID = 3.61$ kips C, $DH = 1$ kip T, $EG = 0$

3-5. $CD = -3.16$ kips C, $IH = 6.0$ kips T, $DI = 3.61$ kips C

3-6. $R_{Ay} = 212.1$ kN, $R_{Ax} = 84.84$ kN, $R_{By} = -212.1$ kN, $R_{Bx} = 0$
$AB = 127.26$ kN C, $AI = 120.0$ kN C, $IJ = 212.1$ kN T
$CD = 21.21$ kN C, $CG = 60,0$ kN C, $GH = 63.63$ kN T

3-7. $R_{Ax} = -25$ kips, $R_{Ay} = -30$ kips, $R_E = 30$ kips
$B_x = 16.67$ kips, $D_x = -16.67$ kips,
On member AC: $C_y = 30$ kips, $C_x = 8.33$ kips
On member EC: $C_y = -30$ kips, $C_x = -8.33$ kips

CHAPTER 4

4-1. $f_v = 27.2$ ksi

4-2.   a. $\Delta l = 0.05$ inches
b. $100,000$ psi

4-3. $\Delta l = .10$ inches

4-4. $\Delta l = .08$ inches

4-5. $P = 120.8$ kips

CHAPTER 5

5-1. $R_{Ay} = 36$ kN, $R_B = 12$ kN, $M = 72$ kN-m @ 2m

5-2. $R_{Ay} = 466.67$ lbs, $R_B = 533.33$ lbs, $M = 4,666.67$ ft-lbs @ 10 ft,
$M = 2,666.67$ ft-lbs @ 25 ft

5-3. $R_{Ay} = R_B = 14.25$ kips, $M = 67.69$ kip-ft @ 9.5 ft

5-4. $R_{Ay} = 25.2$ kN, $R_B = 10.8$ kN, $M = 26.46$ kN-m @ 2.1 m,
$M = 21.6$ kN-m @ 3 m

5-5. $R_A = 7,500$ lbs, $R_{By} = 4,500$ lbs, $M = -7,200$ ft-lbs @ 6 ft, 25, 312.5 ft-lbs @
18.75 ft

5-6. $R_A = 22.25$ kips, $R_{By} = 10.75$ kips, $M = 123.77$ kip-ft @ 11.125 ft, $M = 115.50$ kip-ft
@ 14 ft, $M = 75.25$ kip-ft @ 21 ft

5-7. $R_{Ay} = 8.0$ kips, $R_B = 11.50$ kips, $M = 32.0$ kip-ft @ 8.0 ft, $M = 19.5$ @ 13 ft,
$M = -45.5$ kip-ft @ 26 ft

5-8. $R_{Ay} = 50$ kN, $M_{RA} = 100$ kN-m, $M = 0$ @ 2 m

5-9. $R_{Ay} = 9$ kips, $M_{RA} = 27$ kip-ft, $M = 0$ @ 6 ft

5-10. $R_{Ay} = 180$ kN, $R_B = 90$ kN, $V = 0$ @ 2.52 m, $M = 207.85$ kN-m @ 2.52 m

CHAPTER 6

6-1. $\bar{y} = 8.3$ inches

6-2. $\bar{y} = 34.7$ mm, $\bar{x} = 19$ mm

6-3. $\bar{y} = 11.2$ inches

6-4. $\bar{y} = 7.0$ inches, $\bar{x} = 1.75$ inches

6-5. $I_x = 652$ in^4, $I_y = 74.5$ in^4

6-6. $I_x = 28\,282$ mm^4

6-7. $I_x = 146$ in^4

6-8. $I_x = 1\,5,750$ cm^4

CHAPTER 7

7-1. $f_b = 960$ psi

7-2. $P = 995.56$ lbs

7-3. $f_v = 3.34$ ksi @ N.A., $f_v = 3.13$ ksi @ the top of the web

7-4. $f_v = 75$ psi

7-5. $\Delta = 0.71$ inches

7-6. $\Delta_{T\,\text{Allowable}} = 1.50$ in, $\Delta_{L\,\text{Allowable}} = 1.00$ in, $\Delta_{T\,\text{Actual}} = 1.32$ in, $\Delta_{L\,\text{Actual}} = 0.83$ in

7-7. $\Delta_{\text{Approx}} = 0.41$ in

7-8. 5 inch $\times$ 13 inch beam

7-9. 3 inch $\times$ 12 inch beam

7-10. 6 inch $\times$ 14 inch beam

CHAPTER 8

8-1. 881.25 psi

8-2. 90 kips

8-3. 5,828.1 lbs

8-4. 11,137.6 lbs

8-5. 26.30 ksi

8-6. 74.2 inches

CHAPTER 9

9-1. 2 inch $\times$ 10 inch

9-2. 5 inch $\times$ 12 inch

9-3. 4 inch $\times$ 12 inch

9-4. 8 inch $\times$ 14 inch

9-5. $P_{cr} = 13.27$ kips, $P = 9.35$ kips

INDEX